CORPORATE SUSTAINABILITY

T0265199

This book focuses on corporate sustainability and how it evolves through innovation and new business models. Despite what has been accomplished to date, there is an urgent need for further steps to be taken and this book presents a nuanced but compelling plea for collaboration between businesses, government and civil society.

Drawing upon empirical research, the authors look at recent approaches to corporate sustainability, the circular economy and strategic corporate social responsibility. The book examines these issues from multiple viewpoints, including cultural, social and religious. More specifically, the book explores the freight sector (smart freight leadership), the banking sector (sustainable banking) and Islamic finance and sustainability, detailing the contribution of faith-based organizations to promoting sustainability and the greening of church buildings. Overall, this book captures the emerging new business models and capabilities firms need to implement sustainability.

This book will be of great relevance to students, scholars and professionals with an interest in corporate sustainability, social responsibility, environmental management and eco-innovation.

Jan Jaap Bouma is a consultant in strategic sustainability, addressing the challenges that corporations, governments and non-governmental organizations face in this area. Until recently, he was Professor of Valuation and Governance of Natural Resources at Erasmus University, Rotterdam, The Netherlands.

Teun Wolters is Professor of Corporate Sustainability at Wittenborg University of Applied Sciences, Apeldoorn, The Netherlands.

Routledge Research in Sustainability and Business

Contemporary Developments in Green Human Resource Management Research
Towards Sustainability in Action?
Douglas W.S. Renwick

Environmental Certification for Organizations and Products
Management Approaches and Operational Tools
Edited by Tiberio Daddi, Fabio Iraldo and Francesco Testa

Energy Law and the Sustainable Company
Innovation and Corporate Social Responsibility
Patricia Park and Duncan Park

Corporate Social Responsibility and Natural Resource Conflict
Kylie McKenna

Corporate Responsibility and Sustainable Development
Exploring the Nexus of Private and Public Interests
Edited by Lez Rayman-Bacchus and Philip R. Walsh

Sustainable Entrepreneurship and Social Innovation
Edited by Katerina Nicolopoulou, Mine Karatas-Ozkan, Frank Janssen and John Jermier

Corporate Social Responsibility, Human Rights and the Law
Stéphanie Bijlmakers

Corporate Sustainability
The Next Steps Towards a Sustainable World
Jan Jaap Bouma and Teun Wolters

For more information about this series, please visit: www.routledge.com/Routledge-Research-in-Sustainability-and-Business/book-series/RRSB.

CORPORATE SUSTAINABILITY

The Next Steps Towards a Sustainable World

Edited by
Jan Jaap Bouma
and Teun Wolters

LONDON AND NEW YORK

from Routledge

First published 2019 by Routledge
2 Park Square, Milton Park, Abingdon, Oxon OX14 4RN

and by Routledge
711 Third Avenue, New York, NY 10017

Routledge is an imprint of the Taylor & Francis Group, an informa business

British Library Cataloguing-in-Publication Data
A catalogue record for this book is available from the British Library

Library of Congress Cataloging-in-Publication Data
Names: Bouma, Jan Jaap, editor. | Wolters, Teun, editor.
Title: Corporate sustainability : the next steps towards a sustainable world / [edited by] Jan Jaap Bouma and Teun Wolters.
Description: Abingdon, Oxon ; New York, NY : Routledge, 2019.
Series: Routledge research in sustainability and business.
Identifiers: LCCN 2018015107| ISBN 9781138193758 (hardback) | ISBN 9781138193765 (pbk.) | ISBN 9781315639185 (ebook)
Subjects: LCSH: Corporations—Environmental aspects. | Corporations—Social aspects. | Management—Environmental aspects. | Sustainable development. | Social responsibility of business.
Classification: LCC HD2731 .C64155 2019 | DDC 658.4/08—dc23
LC record available at https://lccn.loc.gov/2018015107

ISBN: 978-1-138-19375-8 (hbk)
ISBN: 978-1-138-19376-5 (pbk)
ISBN: 978-1-315-63918-5 (ebk)

Typeset in Bembo
by Florence Production Ltd, Stoodleigh, Devon, UK

Printed and bound by CPI Group (UK) Ltd, Croydon, CR0 4YY

CONTENTS

FIGURES

TABLES

BOXES

ABOUT THE AUTHORS

Muhammad Ashfaq is the CEO of Amanah Institute of Islamic Finance and Economics, with offices in Germany and Pakistan. He is widely recognized as an Islamic finance expert with more than 13 years of international experience. He has contributed to the development of the Islamic financial industry through research, industry-specific trainings and global initiatives for human capital development and organizational capacity building. He has spoken on the topic of Islamic banking and finance at conferences in more than 30 countries worldwide, and has served as a peer-review referee for a number of international conferences and journals. In pursuit of his passion, he is currently a visiting faculty member of the international MBA programme at Wittenborg University of Applied Sciences in Apeldoorn (The Netherlands), where he regularly teaches courses in finance and management. He is a visiting lecturer in Islamic finance at Coburg University, Germany. Muhammad Ashfaq earned a PhD from the University of Tübingen (Germany). He holds an MBA in Financial Management from Coburg University (Germany), a Postgraduate Diploma in Islamic Banking and Finance and an MBA in Banking and Finance.

Floor Bollee is Manager Communication at Smart Freight Centre (SFC) in Amsterdam, a mission-driven organization for a more efficient and environmentally sustainable global freight sector. She oversees SFC's branding and communication strategy directed at companies involved in freight and logistics as well as government and civil society stakeholders. Floor plays a key role in making complex concepts understandable to SFC's target audiences, both verbally and visually. She has over 15 years' experience in communication, graphic design and advertising with various firms in The Netherlands. She has a bachelor's degree in Sociology from the University of Amsterdam (The Netherlands) and a diploma in arts, specializing in graphic design.

Jan Jaap Bouma is a consultant in strategic sustainability, addressing the challenges faced by corporations, governments and non-governmental organizations in this area. Until recently he was Professor of Valuation and Governance of Natural Resources at Erasmus University (Rotterdam, The Netherlands). He is a supervisor associated with the Off-Campus PhD Program on Cleaner Production, Cleaner Products, Industrial Ecology and Sustainability (Erasmus University, Rotterdam). In his research, he addresses sustainability accounting and strategic corporate management. As an advisor, he deals with issues such as involvement of stakeholders in strategic decision-making processes, in particular regarding sustainability in regional development and the vitalization of industrial areas.

Ankeneel Breuning studied Civil Engineering at the University of Twente (Enschede, The Netherlands), where she obtained a bachelor's and a master's degree. Her master's degree focused on sustainability, with a final project in sustainable building. Currently, Ankeneel is working with ZON Transitie Support B.V. in Ede. This is a company that facilitates the energy transition, aiming for a responsible and sustainable future.

Gizem Goren has been a project finance banker for over 10 years, currently working with Credit Europe Bank NV, Amsterdam. The financing of various sustainable energy projects all around Turkey has increased her knowledge on the implications of sustainability. She enthuses over the fact that funding can make a difference in driving change and progress towards a sustainable future. To provide her experience with a deeper theoretical underpinning, she completed her MBA study at Witterborg University of Applied Sciences in Apeldoorn (The Netherlands) with a thesis on sustainable banking. Gizem is pleased to witness the integration of sustainability and social responsibility as core values in the banking sector. She is dedicated to work on the sector's further progress towards this goal.

Sophie Punte is the Executive Director of the Smart Freight Centre (SFC) in Amsterdam, which she founded in 2013 as a mission-driven organization to catalyse more efficient and environmentally sustainable freight at a global scale. She works with companies that aim to be leaders in sustainability, which involves the monitoring and reduction of emissions across their global logistics chains. She played a leading role in establishing the industry-backed Global Logistics Emissions Council (GLEC) to develop a standard method for emissions measurement, led by SFC. Sophie previously worked at Clean Air Asia, the United Nations, KPMG and an engineering firm on environmental management and corporate sustainability. She is a member of the steering group of the UN-led Global Green Freight Action Plan and World Economic Forum's Global Future Council on Mobility. She has a master's degree in Biology and Environmental Management.

Gilbert Silvius is a Professor of Project and Programme Management at LOI University of Applied Sciences in The Netherlands, visiting professor at the University of Johannesburg in South Africa and fellow at Turku University of Applied Sciences in Finland. He initiated and developed the first MSc programme in Project Management in The Netherlands and is considered a leading expert on the integration of the concepts of sustainability into project management. Gilbert has published widely by means of academic papers and books, and is a frequent presenter on professional and academic conferences. He holds a PhD degree in Information Sciences from Utrecht University, The Netherlands. As a practitioner, Gilbert has over 20 years' experience in organizational change and IT projects and is a member of the international 'enable2change network' of project management experts.

Joyce Forbes Stubblefield is an adjunct professor of environmental sustainability at Southern Methodist University, Dallas, Texas (USA). She teaches two graduate courses: Sustainability Leadership and Earth Matters. She is currently a doctoral candidate at Erasmus University in Rotterdam, The Netherlands, with a focus on sustainability leadership and religious organizations. Her education includes degrees in Mathematics, Business Administration and Sustainability and Development. She maintains her certifications in environmental management systems and Leadership for Energy and Environmental Design (AKA LEED) to promote a more sustainable and equitable future.

Teun Wolters is a Professor of Corporate Sustainability at Wittenborg University of Applied Sciences in Apeldoorn, The Netherlands. He holds a PhD degree in Economics from VU University (Amsterdam, The Netherlands). In the past, he worked with various renowned research institutes in The Netherlands such as TNO, EIM and Statistics Netherlands, carrying out research and contributing to publications in areas such as labour studies, conflict resolutions, changes in long-term health care, sustainability by closing the loops, environmental innovation and environmental management accounting. Moreover, he conducted several projects geared at improving the income and lives of small farmers in Africa. Recently, he initiated the Diaconal Research Centre Apeldoorn (Apeldoorn, The Netherlands) focusing on research that will support the sustainability of local communities and economies.

The views in this book do not necessarily represent those of any of the organizations the authors are associated with, either as employee or otherwise.

PREFACE

This book has a previous history regarding collaboration between the editors. In the late 1990s, we became involved in a European research project on environmental management accounting (called ECOMAC). The project was about obtaining evidence of the extent to which the larger companies in various EU countries had already been active in this field. Moreover, efforts were made to develop the subject further, both by extending existing accounting frameworks and highlighting new approaches. As a spin-off, we participated in the start-up of the Environmental Management Accounting Network (EMAN). Over the course of time this network has broadened internationally; it is very much alive, however, without our being directly involved. Its title is now Environmental and Sustainability Management Accounting (but still carrying the same acronym). We now notice that current researchers still make use of the publications that came to light under the umbrella of the ECOMAC project and adjoining activities.

Since the 1990s, sustainability research has grown in importance by the higher priorities that both society and business are now attaching to sustainability and climate change issues. Despite all the progress made, though, there is still a growing awareness that much more is needed to secure a sustainable and flourishing world for new generations. We live in hope and fear. As researchers, we are looking for ways to make sustainability part and parcel of all economic activities and integrate it into corporate strategies, business models and accounting frameworks. This broader, strategic approach is also present in this book. Indeed, rather than just following the previous path, sticking to a sort of reductionist focus on accounting, we feel that we had to broaden the area by considering the entire complex of value creation, performance and responsible management. Therefore, we, for instance, have included a chapter on freight, a sector that until now was frequently omitted in targeting sustainability. Here, by the way, sound accounting can play a significant role in rendering transparent how seriously freight is contributing to CO_2 emissions and how relevant it is for companies to reduce this effect and opt for sustainable solutions. Moreover, the book contains a chapter on sustainable banking. The banking sector, despite its frequently biased concerns and doubtful moral reputation, cannot be missed when transforming the present economy and creating an inclusive, sustainable society. The chapter on sustainable banking shows what a commercial bank has to live up to in order to qualify for the title of a sustainable bank.

Both for people and companies it seems to be very difficult to take a longer view and base one's action on it. We tend to be guided by short-term needs and convenient arrangements. The time of writing of this preface coincides with the Dutch 'week of books'. On this occasion, the Dutch former politician Jan Terlouw (now in his eighties) wrote an essay on the alarming decline in biodiversity and the threats it entails. He highlighted human behaviour by the following comparison. When a Dutch bridge showed some hairline cracks, it was immediately closed so that the traffic normally using it had to make a detour of 50 km. However, when, at the beginning of the 1990s Shell produced a video which, based on sound analysis, showed that the earth's temperature would increase with 4 degrees Celsius by 2050 if climate policies remained unaltered. But then, nothing changed. He continued by arguing that, obviously, we as humans have no innate long-term gene. Therefore, such a 'gene' has to be developed based on our excellent analytical brains. We need nature to breeze, to eat, to be part of it. His overall message is: we can do if we want to. This book is an attempt to say the same in terms of corporate responsibility.

Jan Jaap Bouma and Teun Wolters

1

INTRODUCTION

Corporate sustainability: requirements, realities and prospects

Jan Jaap Bouma and Teun Wolters

Corporate sustainability

Over recent decades, corporate sustainability and related concepts such as Corporate Social Responsibility (CSR) have gained in importance. Many companies have adopted policies that are based on them.

This book is dedicated to breaking new ground and following new paths. Many people who are aware of the importance of working on a sustainable world will welcome the growing attention that corporate sustainability is receiving. However, this book is not primarily written to applaud the state of affairs, but to help companies and others pave a path to the future.

Beside positive action and praise, there has been criticism of corporate sustainability, which can be a matter of either casting doubt on the intentions of companies (business as usual or almost as usual) or a critical consequentialist appraisal of what present-day corporate sustainability has achieved. To make a longer story short, as it stands now, there is underperformance, no matter how one looks at it. The dissatisfaction that follows is often invested in pleas for real transformation or a new paradigm. Corporate sustainability based on such relatively radical notions has been worked out in different publications and programmes. However, seemingly, there seems to be a lack of leadership needed to implement it broadly.

An evolutionary process

How one judges this predicament can be influenced by whether one accepts that companies have to go through an evolutionary process that starts with a rather defensive attitude and proceeds towards a proactive resolve to integrate sustainability in the company's core business. This book takes that position. That does not mean the process is going fast enough. 'Too little and too late' often fits well with the progress made, at times evoking moments of cynicism or even despair.

To speeds things up and enhance sustainability strategies, this book discusses how corporate sustainability becomes embedded in wider processes of societal change and innovation. Interestingly, this involves both a focus on individuals, particularly senior staff members who are revealed to be leaders in sustainability and have an eye for the wider society. The latter can be a matter of co-evolution (society preparing for the adoption of new priorities and radical innovations) or a matter of selecting and associating the company with like-minded parties (but with complementary competences) to create inspiring collaborative innovation networks.

Sustainable value placed in a wider context

Today, both kinds of societal involvement are occurring in many places. The overall aim is to create sustainable value. For this to happen, there is a need for imaginative entrepreneurship, solid business planning and various forms of community-based collaboration. In the quest for sustainability, there is a place for:

- big business taking responsibility
- SMEs seeing the niches
- small-scale local endeavours
- central governments issuing adequate programmes and budgets
- local authorities working together with civilians to get things off the ground
- individual positive mindsets and inspirational leadership
- solid monitoring and controls.

Within this wide pallet of contributing parties, there is no need to see a contradiction between, on the one hand, the development of business-like Key Performance Indicators (KPIs) to monitor what is going on and the values and attitudes needed to be motivated and persevere, on the other. Both are required. Sustainability can be enhanced as it is motivated by the three constituent factors: external pressures, internal motivation and opportunities. There will be individual differences in emphasis but, where these factors reinforce each other, sustainability in society can move to the next level. Project management deserves special attention as many corporate decisions are implemented by means of special projects. It is of crucial importance that a company's projects are executed in line with the existing corporate sustainability objectives. Project managers should not be fully dependant on those who commission a project when reaching established sustainability objectives is at stake; instead, they deserve a degree of autonomy when sustainability is at risk.

A special point is the need for broadening the scope of corporate sustainability. This book explores this area by means of two chapters focusing on a particular industry. The first of these is Chapter 5, on smart freight leadership. For a long time, the freight and logistics sector has been under the radar when it came to environmental monitoring and serious environmental policies. The chapter shows how that can be changed, making a convincing case for harmonizing the measurement of GHG emissions on which leadership in sustainability can be based.

The other chapter devoted to a particular industry is Chapter 6, Sustainable Banking. This chapter applies a model that shows the various phases traversed by a company on its way to becoming a fully integrated sustainable business. The chapter offers a remarkable insight into what this means in terms of ditching wavering policies and reaching crucial tipping points.

This book also pays attention to the influential role that religion can play in motivating people to be serious about nature and the environment. Part of this includes religiously inspired initiatives in the area of ethical trading. Chapter 7, on Islamic finance, shows how the ethical foundation of Islamic finance (banking and insurance) can inspire care for the social and ecological aspects of human activity. However, that effect can only be optimal if the ethical foundations, formulated a long time ago, keep up with the times. For instance, the concept of stewardship (present in both Christianity and the Islam) can be inspiring in pursuing sustainable practices only if it is updated in order to incorporate contemporary issues. Chapter 8, on faith-based organizations, shows how in the past such organizations have inspired key social activities while today these cannot be overlooked when considering what they contribute to areas relevant to implementing sustainability. It appears that the various religious sectors based in social peace, justice, equity, development, and stewardship also characterize, capture, and articulate sustainable acumen throughout its global and local networks. Part of this is a focus on the greening of churches and other religious buildings (Chapter 9). This chapter develops a model for this that not only facilitates the technical measurement of an ecclesiastical building's energy use, but also includes the responsibility of the religious community to optimize the building's energy use and, where possible, adopt sustainable energy sources.

Overview of the chapters

This section gives an overview of the various chapters of the book by providing brief abstracts.

Chapter 2. Sustainable value creation

This chapter argues that many organizations have begun to make sustainability a part of their strategies and innovation efforts. Because of sustainability, the value to be created by businesses needs to go beyond business-as-usual. Various concepts that can help to make sustainable value creation concrete are discussed, such as Corporate Social Responsibility, shared value, the sustainable city, the sharing economy and the circular economy. The chapter also goes into the contribution of innovation to sustainability by discussing the concepts of responsible innovation, social innovation and sustainability innovation. The chapter is rounded off by asking whether today's business schools are sufficiently preparing our future managers for the transformation that is to take place. Millennials appear visibly interested in responsible and purpose-driven businesses. A future-oriented approach in business curricula is within reach.

Chapter 3. Sustainable strategies and accounting needs

Many companies have introduced strategies and practices that relate to sustainability. That explains the continuous attention in research given to the development of new strategies and related business models in which sustainability is given a prominent place. Consequently, there is also research in the area of management accounting that can support the strategies and management control in the area of corporate sustainability. This chapter explores various avenues to promote sustainability as a major strategic management issue and how these call for supporting accounting information. Special attention is given to the following areas:

- business continuity and environmental management
- corporate governance and sustainability
- responsible Management
- hyper-competitive markets, strategies and organizational adaptation
- business models and sustainability
- the rise of community-based strategies, innovation and management accounting.

Chapter 4. Sustainability in project management: making it happen

Based on the available literature, this chapter identifies the areas of impact of sustainability on project management such as selection of the project team, context, stakeholders, project objectives and specifications, dimensions of project success and communication. The chapter emphasizes the pivotal role of the project manager in realizing an organization's sustainability; in this, the project manager not only takes orders but accepts special responsibility for the inclusion of sustainability in the projects he or she manages. Sustainable project management involves managing a project's social, environmental and economic impacts by an approach that considers flexibility, complexity and opportunity.

Chapter 5. Towards smart freight leadership

Smart or green freight represents the efforts to transform the freight and logistics sector in order to reduce its greenhouse gas emissions and air pollution. This is to be achieved by improving its fuel efficiency across the global logistics supply chain without mitigating its vital economic functions.

Multinational firms and their brands undergo growing pressure to reduce their carbon footprint. This market pressure cascades down the global supply chain. Green freight programmes respond to these business needs.

The chapter emphasizes that the world needs genuine smart freight leadership. Studies show that only a fraction of cargo owners takes up that role. A major barrier is a lack of standard methodologies to calculate their carbon footprint and define emission reduction targets.

To create and implement a universal method of calculating logistics emissions, the Smart Freight Centre established the Global Logistics Emissions Framework (GLEC). If this framework is accepted by industry, governments and NGOs, it will be possible to harmonize the various green freight programmes around the world and make it possible to measure and compare their effects.

Chapter 6. Sustainable banking: the mysterious role of commercial banks in achieving a sustainable economy

In the wake of the economic crisis of 2008, conventional banking seems to be giving a higher priority to sustainable policies and practices. A new generation of bankers called for a serving, diverse, outspoken and sustainable banking sector. Many banks have made some progress in this area but the task is certainly not finished. To enlighten this mysterious area, this chapter focuses on how a bank can be evaluated and positioned in terms of corporate sustainability.

To address this area, this chapter uses a phase model of corporate sustainability. This model describes various phases through which companies traverse in the transition towards corporate sustainability. A major part of the research is about adapting and applying this model to the commercial banking sector. Moreover, the model was applied to a major commercial bank in The Netherlands, leading to interesting observations and conclusions.

Chapter 7. Islamic finance and sustainability

This chapter discusses Islamic finance and how it can contribute to sustainability. The chapter connects Islamic finance with what Islam teaches us about the world and how to sustain it. Islamic finance can be considered as a framework for a type of banking and insurance that meets certain criteria that may be helpful in promoting sustainability. However, to strengthen the bond between Islamic finance and sustainability, Social Responsible Investment (SRI) could be a useful reference rather than focusing on conventional banking and finance and how to emulate them.

Chapter 8. Faith-based organizations and corporate sustainability

Faith-based organizations (FBOs), along with forms of 'religious capital', appear to have a role in corporate sustainability. This role is contextually bounded and manifests itself through leadership, responsible investments and dialogues within and between different communities. Various entrepreneurs accept FBOs as community problem solvers. From the 1970s to the present, religious organizations' economic approaches and social networks have played key roles in implementing and managing sustainability-led projects by individuals and organizations. Methodist and Quaker immigrants introduced the concept of social responsibility in investing to the USA; for more than 200 years, this faith-based practice is still extant. In a broader context, large religious organizations with billions of adherents are facing membership challenges in addressing societal ills. In particular cases, religious leaders have to take responsibility for building stewardship by means of sustainable business acumen involving investments that are beneficial to both the environment and society at large.

Chapter 9. The greening of churches: a maturity framework based on policies and measures

Improving the sustainability of a church building can be achieved by a framework of maturity levels. This chapter presents such a framework and its application to three churches in The Netherlands. The maturity levels provide insight into the current level of sustainability policies and measures, but also into where opportunities for improvements lie. The framework contains a questionnaire that allows a scoring for an individual congregation and their church building. It is a starting point for discussions on what it means to be a green church.

Concluding remarks

After all, one may raise the question: do all the strategies, business models and dedicated accounting tools that are aimed at promoting corporate sustainability eventually lead us in the right direction and ensure that a sustainable future is within reach? No definitive answer is available, but this book opts for a hopeful 'yes'.

With growing pressures and newly emerging opportunities, it is necessary to embrace more radical solutions. However, this is part of an evolutionary process that unsettles many things without making it possible to give up all conventional approaches that have shown their limitations. For instance, good planning and monitoring still have their place. These have advantages compared with a situation packed with good intentions but without proper and controllable action plans. This, nonetheless, goes hand-in-hand with doubts about whether planning and control can make a real difference in terms of sustainability. In their planning, companies continue to deal with consumers that are conditioned by unsustainable practices while companies paradoxically depend on them for their survival. This may go back to the concept of sustainable development that, after all, supports economic growth, for sure, with more equity and a better care for the environment; but still, it is about economic growth. This leads to a situation where, on the one hand, companies and governments are comforted by the thought that sustainability can go hand-in-hand with economic growth and job creation (the 'green economy') and on the other hand, are called upon to create prosperity without growth (Jackson, 2010).

The neoliberal wave of past decades has indeed led to a great deal of economic growth, but this triumph of international trade came with considerable inequality, economic crisis, environmental degradation and unwanted climate change. In various countries, this development has led to political confrontations and anti-liberal forms of democracy and anti-humanitarian populism (Crouch, 2017). Decentralized forms of government can be in the interests of sustainability, but at the same time there is a great need of international leadership by nations, international organizations and international companies that show the path to a sustainable world. There are already signals of such leadership . . .

A world in transition creates both hope and fear. This mixture of emotions also signals the complexity of today's reality. Uncertainty may be a matter of knowing what is unknown, allowing us to define at least some scenarios about the future. However, complexity implies that we don't know what we don't know. This is both scary (you'll never know what is going to happen) and hopeful (new, unexpected developments for the good can emerge). What is important is the realization that the flows of optimism and pessimism created by a complex world in danger require a special kind of leadership. The experience of *change per se* and the dissonance it creates fuel new thinking, discoveries and innovations based on leadership (which can be rather informal) that does not dictate but, in conjunction with others, creates sustainable values (Ferdig, 2007). The more-than-problems that need to be addressed in a complex world cannot be solved by conventional approaches (along the usual left–right divisions). Caring and a degree of pragmatism away from narrow-mindedness (based on ideological thinking) is what the world needs now (Ehrenfeld and Hoffman, 2013).

As becomes evident today, rather than technology it is human behaviour that is the main bottleneck in reaching credible levels of sustainability. Nonetheless, over a longer period of time, futurologists may see trends of technological developments that go through waves of behavioural adaptation and make it possible to enter a sustainable world. An example of this is what Rifkin (2010) calls The Third Industrial Revolution, which will dramatically change the globalization process. To make a longer story short, this is based on the spread of renewable forms of energy – solar, wind, hydro, geothermal, ocean waves and biomass – and is one of the pillars of the Third Industrial Revolution. Loading and storage technologies will be crucial in this process (hydrogen is expected to make this possible). The commercial and economic implications are far-reaching for the real-estate industry, and for that matter the world. If millions of individuals in developing communities were to become producers of their own energy, the result would be a notable shift in power relations. Communities would be able to produce goods and services locally and sell them globally. This is what Rifkin sees as the essence of the politics of sustainable development and re-globalization from the bottom up.

This longer-term technology-oriented vision, however, can be reliable only if such processes are embedded in humane governance structures based on human rights and development perspectives based on power sharing and the peaceful settlement of disputes. Is this too far-fetched?

References

Crouch, C. (2017) *Can Neoliberalism Be Saved from Itself?* London: Social Europe Edition.
Ehrenfeld, J.R. and Hoffman, A.J. (2013) *Flourishing. A frank conversation about sustainability.* Stanford, CA: Stanford Business Books.
Ferdig, M.A. (2007) Sustainability leadership: Co-creating a sustainable future', *Journal of Change Management*, 7(1), pp. 25–35.
Jackson, T. (2010) *Prosperity without Growth: Economics for a finite planet.* London: EarthScan.
Rifkin, J. (2010) *The Emphatic Civilization: The race to global consciousness in a world in crisis.* Cambridge, UK: Polity Press.

2

SUSTAINABLE VALUE CREATION

Teun Wolters

Individual and collective responsibility

Creating sustainable value

If the purpose of economic activity is the creation of economic value, then it is obvious that sustainability requires the creation of economic value that is sustainable. As long as sustainability is not normal procedure, a distinction between economic value and sustainable economic value remains relevant and even crucial. In fact, the entire issue of building an economy that is future-proof – i.e. that does not restrain future generations from living a worthy and affluent life – goes back to this distinction.

Today's concerns about a sustainable world have led to a broad terminology to express its various dimensions and implications. Sustainable development, often simply called sustainability, could be the overarching term to cover all the economic, social, ecological and governance dimensions of creating a future-proof global human society. Others nominate 'corporate social responsibility' or 'responsible management' as the term that unites all dimensions.

Going by the Principles of Responsible Management Education (PRME), there has been a shift from a traditional organizational focus on what businesses should be and do to a focus on the manager as a person (what kind of person should a manager be and what should he/she do and how?). It suggests that emphasizing the managerial and operational achievement by a responsible manager could be the 'logical next evolutionary step' of translating the organizational vision on sustainability (Laasch and Conaway, 2015).

Indeed, the individual perspective is important and even crucial when it comes to daring strategic choices and various forms of entrepreneurship needed to make an organization more sustainable (or more responsible, if you like). However, the personal dimension should also mean that individual managers and employees should take responsibility for another sustainability requisite – that is, going beyond the boundaries of a single company, involving strategic alliances, supply chains, communities or networks. It is also important to realize that sustainability is to become part and parcel of an organization's culture, which implies a kind of collective responsibility for creating and enhancing a business's sustainability.

There is much to be said in favour of choosing sustainability as the all-embracing term for all dimensions of creating a tenable future for mankind – not in the least as it keeps reminding us of the future and future generations. Nonetheless, given the relevant literature, other terms, such as corporate social responsibility (CSR), could play a similar role.

Different theoretical concepts have played a prominent role in making people aware of the sustainability imperative. Two concepts of this kind will be discussed here, external effects and shared value.

External effects

In neoclassical economics, it is understood that a market transaction may also have notable effects on others who are not engaged in that transaction. These effects are called external effects (or externalities). They can be either positive or negative. Environmental harm inflicted on third parties resulting from economic exchange transactions has become a major example of a negative external effect.

Until the 20th century, economists did not explicitly recognize external effects. When a few economists began to consider these a century ago, they assumed that external effects were of minor importance (Ayers, 2014). The famous economist Marshall made a first step in recognizing the existence of external effects by analysing situations of 'unpriced scarcity', but he did not consider the cost implications of such effects. This was elaborated on by Pigou, who suggested that the acts of market parties also influence the prosperity of others not involved in a deal. He mentioned, for example, the negative external effect of the smoke emitted by a factory: '. . . it inflicts heavy uncharged loss on the community' (Crane *et al.*, 2014). Joseph Schumpeter introduced the notion of 'spill-overs' from business innovations, which in turn depend on investment in education or R&D (Schumpeter, 1934). These are mostly forms of positive external effects. However, as modern forms of production continued to spread, negative external effects implied serious damage to masses of people and the environment. In the 1960s, Rachel Carson's book *Silent Spring* revealed how pesticides used by farmers to kill harmful insects also killed the birds that ate the insect eggs and grubs (Carson, 1962). One of the probably most dangerous external effects is the increasing concentration of greenhouse combustion of fossil fuels (Ayers, 2014). This is expected to lead to a change in climate that will harm people in large parts of the world.

To make a longer story short, neoclassical economics sees the environmental problems that we encounter today as the consequence of market failure. That is, market prices do not sufficiently reflect the scarcity of natural resources and environmental sanity (such as clean air). Therefore, many goods and services are priced too low so that they are bought in quantities that do not respect the scarcities of the natural inputs needed to produce them. It seems to make sense to correct prices in a way that makes products that use relatively large proportions of the available natural resources more expensive, and other products that use relatively fewer natural resources cheaper. Although, both theoretically and practically, increasing prices based on insights into the environmental effects of products is not always easy to establish (because of a lack of information or unwanted competitive effects), the idea as such has not disappeared. Taxes and subsidies play an important role in government policies promoting environmental innovations and environmentally friendly products. These policies, however, are highly fragmented and often unsatisfactory. From that point of view, there is not only market failure, but also government failure. For instance, subsidies to influence market decisions can create highly adverse incentives from an environmental point of view. In recent years, there has been an intensified discussion on simultaneously bringing down the costs of labour by reducing income taxes and social charges and increasing the prices of products and services to reflect the environmental resources to which they lay claim.

Shared value

The literature on CSR seems to be on a par with that on sustainability. There is multiple CSR research that deals with the issues that equally belong to the realm of sustainability. Within this context, there is strategic CSR, which is to ensure that CSR is becoming truly integrated in the firm's strategy and its implementation.

Strategic CSR moves beyond good corporate citizenship and mitigating harmful value chain impacts; it tends to focus on a small number of initiatives whose social and business benefits are large and distinctive. Strategic CSR involves both inside-out and outside-in dimensions working in tandem – that is, reaching out to the external environment and involving external stakeholders in internal CSR decisions to develop business that serves simultaneously the interests of different parties, whether directly part of a deal or not. Porter and Kramer (2006) introduced the concept of shared value to highlight what strategic CSR could entail. Strategic CSR unlocks shared value by investing in social context that directly or indirectly strengthens company competitiveness. A symbiotic relationship develops: the success of the company (including its supply chains) and the success of the community become mutually reinforcing.

Most of the time strategic CSR occurs when a company adds a social dimension to its value proposition, making social impact integral to its overall strategy.

Integrating business and social needs takes more than good intentions and strong leadership. It requires adjustments in organization, reporting relationships and incentives. Companies must shift from a fragmented, defensive posture to an integrated, affirmative approach. The focus must move away from an emphasis on image to an emphasis on substance.

According to Porter and Kramer (2006), perceiving social responsibility as building shared value rather than as damage control or as a PR campaign will require dramatically different thinking in business.

Shared value has become a 'mobilizing' concept. It has been vocally embraced by all kinds of organizations (firms, business schools, NGOs, etc.) and has made headway into the academic management literature (Crane *et al.*, 2014). 'Creating shared value' (CSV) is not a fully original concept, but rather is about what the academic literature has put forward on CSR, stakeholder management and social innovation. Moreover, CSV seems to ignore or underestimate the tensions between social and economic goals. There may at times appear important win-win opportunities, but these cannot conceal the many situations where social and economic outcomes are not beneficial for all stakeholders. Even for the firm aiming to act responsibly, CSR does not always imply greater profits (Crane *et al.*, 2014). CSV builds on the assumption that compliance with legal and moral standards can be expected to happen. Under such a condition, governments can play a positive role in ensuring the success of CSV by regulating what must be done and avoided. In fact, this is happening is many countries and places. However, especially in the case of multinational corporations, which operate in a broad variety of geopolitical contexts, governments frequently appear unable or reluctant to play such a role. At a closer look, CSV seems to meet its limits in a continued adherence to a rather narrow economic view of the firm (competition, short-term profits, conventional financial criteria) despite a professed stakeholder approach and a recognition of the firm as a multi-purpose entity (Crane *et al.*, 2014).

By way of an intermezzo, one may wonder how serious this is: does this discredit the CSV concept? It is safe to say that all views on what policies we need have their limitations; they may inspire new ways of practising corporate sustainability but, especially as public expectations are rising, may also lead to new forms of window-dressing and green-washing. The latter are a shame, but at the same time ignite the engine that elevates society to the next levels of sustainability. We are living in a dynamic world. That means, amidst failure and worries about how the world is developing, new windows of opportunity appear, opening spaces for new ideas and new business models that bring sustainability closer.[1]

We see a gradual but clear development in the thinking of CSR. Where, as in neoclassical economics, one's point of departure was a given production technology (production function) and a given market (demand and supply functions) with a clear legal framework as the only peripheral condition, a firm had just one social responsibility – that is, making the best out of it by choosing the production volume that maximizes profits. However, reality can be quite different. Especially in a dynamic global economy, there is much uncertainty because of limited information, new technologies, new legislation, changing political forces, changing consumer tastes and fashions, and broadening global markets. These factors require strategic priorities and well-considered business models (involving certain time horizons) while defining one's ethical priorities and responsibilities towards one's key internal and external stakeholders. All these things together make up a firm's value propositions towards its customers and society. From there, it can be assessed how and where sustainable value creation is on the horizon.

To position itself in terms of social responsibility, a firm must articulate the following (Carroll, 1979):

1 A definition of social responsibility (e.g. does it go beyond economic and legal concerns?).
2 The issues for which a social responsibility exists (depending on the industry, e.g. environment, product safety, human rights, local employment).
3 Its philosophy of social responsiveness (its behavioural attitude ranging from no response, reactive, defensive, accommodating to a fully proactive response).
4 The specific social issues involved in the stakeholder relationships: environmental concerns, product safety, human rights issues etc.).

The kind of responsiveness (see 3, above) depends on a firm's understanding of where a specific issue relates to its economic, legal, ethical and other responsibilities.

There are many empirical studies that have examined the relationship between corporate social performance and corporate financial performance. There are many social problems in the world that need to be attended to. If a company takes these, or at least some of them, to heart, does that mean lower profits? And, how important is that? And, what time horizons are to be considered here? If the profit motive remains undifferentiated, and without embedment in a social-responsibility framework, CSR will follow an instrumentalist logic only (Margolis and Walsh, 2003); that means, CSR will be just a (probably useful) tool to better address the vagaries of the market and the rest of the external environment. From this perspective, it is understandable that much research has been done to demonstrate a positive link between CSR and financial performance. Wherever there is a positive link, CSR can be defended as equal or even superior to 'business as usual' without CSR. However, there seems to be an instability in the research outcomes due to a variance in how these studies were carried out. Some results are positive, others negative or neutral. A few researches claim that there is a sort of U-shaped success curve. There is a middle range of unsuccessful efforts. For CSR to boost profits, it seems necessary to perform a full-fledged CSR strategy for the entire company (Galant and Cadez, 2017).

Although profits and CSR are not mutually exclusive, it is essential to recognize there are several normative issues that companies cannot ignore. Such a normative approach to CSR tells the corporate world there are issues such as human rights abuses, environmental issues and poverty that demand the attention of all business communities. Companies are expected to help alleviate these ills. They should consider these issues, at least to the extent that they are entwined with or touch upon their own business. Moreover, companies can be expected to develop products and services that help address social (including ecological) issues if these are close to their technological and organizational capabilities. This is also inherent in the shared-value concept. CSR can be stimulated by influential stakeholders pleading for a company's social response in these areas. However, a company cannot escape basic moral issues, also when stakeholder pressure is absent. In the final analysis, it is the future generations that are the key determining factor (stakeholder) in regard to a sustainable future. Therefore, stakeholder interests and profit calculations should be subject to a quality check: are they convincingly future-oriented? Discussion on strategic CSR is about making profits that take the planet into account – this is what is also called transformational leadership, or transformational business. Current businesses cannot get around the present unsustainable economic system but are challenged to make moves within that system that support the transformation of that system towards a sustainable economy. Where that transformation is not served, there is only business-as-usual, whether done openly or covered up by symbolic policies or forms of green-washing.

How to serve the transformation?

Transforming business models

Companies serving the transformation towards a sustainable future do not operate in isolation. Besides having and further developing their own values, priorities and objectives, companies need both a demanding environment and opportunities for innovation. There are examples of companies that have their beginning in one of the three drivers but, in the final analysis, all three are needed. Sustainability at large requires the prioritization of the basic environmental issues that threaten a liveable planet and how they can be addressed in concrete situations, where possible in coherence with the other normative CSR issues.

Three major business models have been particularly promising in serving the transformation (Henderson, 2015):

1   forestalling risks by preventing brand damage and preserving a 'license to operate'
2   increasing operational efficiency
3   selling to the environmental niche.

The forestalling of risks

An increasing number of concerned consumers, empowered by global media and social networks, have led many firms with large brands to invest in sustainable business practices to prevent brand damage. Also, strict regulatory environments or critical communities have caused companies to invest to head off potential regulation or to prevent the loss of their licence to operate (either literally or figuratively). Examples of companies that have operated in that manner include

Nike, McDonald's, Kimberly Clark and Coca Cola (Henderson, 2015). In some cases, environmental actions at industry level have been motivated by potential regulatory pressures, the chemical industry being a good example. Companies can also be persuaded to consider the long-term viability of their portfolios because of their climatic impact and vulnerability to radical regulatory change. If they fail to do that, they risk being left with worthless assets, useless resources and obsolete business models (The Economist Intelligent Unit, 2017).

Increasing operational efficiency

One of the most direct impacts of the environmental crisis has been rising input costs, often leading to an increase in commodity prices. Therefore, increasing the efficiency by which resources (especially energy) are used is a prominent way to generate economic gains (Henderson, 2015). The same goes for the construction of environmentally benign buildings. The costs of initial construction of building according to new standards may increase by up to 8%, but these represent only 5–10% of total life-cycle costs. The reductions in energy and water over the lifetime of the building lead in many cases to an overwhelming advantage. Technological progress is expected to continue over time, making the saving of inputs increasingly attractive. New firms may emerge making use of saving opportunities right from the beginning while often introducing 'integrated solutions' (Henderson, 2015).

Selling to the environmental niche

A relatively small proportion of consumers appears to be willing to pay more for sustainable products, and there is a sizeable number of companies that focus on such products. Examples include Starbucks, Toyota selling its Prius, and the Triodos Bank (exclusively investing in sustainable projects) (Henderson, 2015).

To some degree, business resting on environmental sustainability is a matter of anticipating a continuous transformation to a fully sustainable economy. Investing in anticipation of that transformation is a risky endeavour. Nonetheless, it is obvious that transformation towards sustainability needs real entrepreneurs who lead the way in both fully commercial and social innovation. This is a dynamic process. It implies that it is not always easy to judge whether a company is on a truly sustainable path. The unsustainable present economy requires a sense of realism in regard to what suppliers and customers will accept as feasible propositions; so, besides anticipation and creativity in designing business models, timing is a crucial element of a sustainability strategy. The dynamic nature of a sustainability strategy may also imply that what was progressive at one point in time could turn out to be outdated at another point. Besides an internal development towards integration of sustainability in a company's core processes, there has to be sufficient alignment with the strategic environment as innovation normally requires collaboration and facilitation by third parties.

The sustainable city is our future

Economic activities tend to concentrate in and around cities. This implies migration to the urban areas. Today, already 50% of the world's population lives in cities; that will be 80% in 2050. Especially in China and Brazil, this development is overwhelming.

Concentration in population will lead to more social conflict and environmental pollution. Moreover, The Netherlands as a lowly situated delta country must take into account flooding and increasing temperatures resulting from climate change. These problems can be used to shape our cities in a sustainable manner. By doing so, we improve the ecological and social living conditions while simultaneously strengthening our economy (Cramer, 2013).

What must be done?

Cities are breeding grounds for renewal, innovation and creativity. This is an important basis for urban development. The most important task is the transition to sustainable energy provision and a circular economy that reuses water and turns waste into input material. Such a transition will have consequences for all economic activities.

New buildings will be substituted by large-scale renovation. Houses, offices, shops and other buildings will have to be adapted to the wishes and requirements of the 21st century. Sustainability is a major factor in this renovation wave. It offers employment and improves ecological and social liveability.

The present energy sector will undergo increasing competition from local energy companies as sustainable energy suppliers. Smart grids will play a major part in this. They coordinate demand and supply in an efficient way. Our building will be future-proof by saving measures and sustainable energy. Our energy bills will reduce, CO_2 emissions will decline and our standard of living will rise.

Table 2.1 shows several generic transformational activities and brief descriptions of what these entail (Cramer, 2013).

The sustainable city concept is an invitation to integrated thinking. That implies avoiding fragmented solutions to problems and avoiding short-termism in costing and budgeting. Instead, we should be thinking in terms of values and opportunities. Renewal can readily be hampered by a separation between developers and administrators, which is common practice in the building industry. Securing a building's value over the course of time needs a realistic view on which investments are to be made, which requires a timely consideration of energy, maintenance and renovations. Doing this may look simple; technologically speaking, almost anything can be realized. However, routines tend to dominate human behaviour so that renewal takes place slowly. New forms of collaboration, based on learning how to deal with opportunities, values and innovation, need to be developed in order to overcome 'old-fashioned ways' (Dagevos et al., 2017).

In many regions, there is mention of an urban renaissance. For instance, during the 1990s, the city of Eindhoven was in a pessimistic mood as major companies

TABLE 2.1 Transformational activities in cities

Activity	Brief description
Renovation	New buildings are substituted by large-scale renovation. Houses, offices, shops and other buildings are adapted to the requirements of the 21st century, sustainability being a major factor. This offers employment opportunities and improves the ecological and social quality of life.
Local sustainable energy	The present energy sector undergoes competition from local energy companies as sustainable energy suppliers. Smart grids play a major in this, helping to save energy. New forms of collaboration and financing emerge.
Recycling	Consumer products are no longer discarded, but reused. New ways of collecting used materials serve reuse and recycling.
The water cycle	The water sector undergoes a metamorphosis. Water is reused locally and stored in special buffer reservoirs. Waste water is a source of energy and is reused (e.g. phosphate).
Cleaner mobility	There is growth in the use of public transport, cycling and taxis, as well as shared means of transport.
Greening	There is more vegetation in the streets and on buildings; this reduces heat and improves their insulation.
Circular economy	Sustainability changes the entire metabolism of cities: the flows of food, waste, water and energy. Rather than a linear process, the economy is evolving towards a circular economy.

Cramer (2013)

disappeared and Philips halved its workforce (from 40,000 to 20,000 employees). The EU's Stimulus programme was used to confront the crisis. Nowadays this kind of pessimism has totally disappeared. Eindhoven is proud to demonstrate its recent achievements. The Brain Port region (belonging to Eindhoven) is considered to be the second engine of the Dutch economy. Being smart is an important element of its image, which suits the city given its prominent high-tech industry. The former industrial parks are transformed into vibrant new areas spearheading economic activity.

However, the increasing economic strength of many cities seems to go together with social deprivation and inequality for certain segments of the population and quarters. Moreover, in several cases, urban prosperity runs parallel with a deterioration in living conditions in rural areas. The process of gentrification and segregation begs several compelling questions as to whether the growing inequalities should be accepted as an unavoidable development or whether these call for a social programme to counterbalance them (Dagevos *et al.*, 2017).

Within the context of creating sustainable value, it is also relevant to realize that getting a picture of a city's economic value creation must include the negatives that take place outside the direct scope of a project or transaction. Also, here, we see the need to simultaneously consider economic, social and environmental sustainability from a wider systemic perspective (Dagevos *et al.*, 2017).

Focusing on people's immediate living environment cannot offer solutions for all ecological problems faced by the earth, but it invites improvement in the immediate living environment in different ways. Possible actions include the planting of more trees or shrubs in the neighbourhood and implementing a multi-functional garden in which residents grow vegetables, herbs and fruits. This garden enterprise leads to healthy exercise, healthy food, better social contacts and an increasing number of social initiatives. Another example is about working on a Blue Zone, where older people are relatively active, healthy and happy. Reducing the concentrations of medicinal residues in surface water was one of the spearheads of such an initiative. Here, a transition from medication to prevention through more physical activity and contact with nature takes centre stage (Staps *et al.*, 2017).

Health as an overall indicator of the quality of life

Creating sustainable value is a continuing story of considering various aspects of life that for a long time remained out of sight or were dealt with in a fragmented fashion. GDP still functions as a key wealth indicator but it is increasingly evident that a model of human development based on economic progress alone is incomplete. The Social Progress Index, for example, is one way of addressing that incompleteness (Porter and Stern, 2017).

Recently, the concept of positive health was launched as a policy compass showing the way to improving people's quality of life. Over recent decades, healthcare costs have increased substantially. This increase is associated with ageing and the availability of a wider range of, often costly, therapies. There is a problem with regarding mounting healthcare costs as an indicator of increasing prosperity. 'Positive health' is the ability of people to adapt and self-manage their lives; it has six domains: self-management of daily life, physical functions, mental well-being, meaning, participation in society and quality of life. The broadening of the concept of health from 'not being ill' to the six domains of positive health could be a first step in connecting sustainable value with health and the human living environment. The positive and negative effects of economic activity in terms of the domains of health could be helpful in breaking away from the conventional, rather one-dimensional, focus on economic growth. The model inspires the definition of interventions that enhance people's health by integrating the various policy domains without necessarily incurring additional financial costs (Staps *et al.*, 2017).

The wider link with sustainability here is that people's way of life, including production and consumption, affects not only health but also the ecological vitality of the earth.

Lageweg (2017) pointed out that there are many aspects that have gone wrong with our food supply. Over 50% of chronic diseases are associated with eating habits and lifestyle. Poor and unhealthy food demonstrably reduces the vitality of both young and old. Also, many behavioural issues derive from certain consumption patterns. Altogether, these issues result in enormous societal costs. At the same time, there has emerged much scientific evidence that healthy, fresh and varied foodstuffs make a major contribution to both the prevention and cure of diseases. For instance, by changing eating patterns, the use of diabetes medicines can be markedly reduced. Moreover, projects with schoolchildren, the elderly, the sick and prisoners have shown that healthy and varied food has a marked effect on how people behave and act. Food markets, through many links, are dominated by a limited number of big players opting for large quantities against the lowest possible price. In this system, nutritional value and sustainability are barely rewarded. Because of that, many farmers and workers in the food chains have low incomes and poor livelihoods. Among both farmers and consumers, there is a counter-movement emerging. Concepts such as organic, regional products and slow food are gaining ground. Various companies, such as store chains, have adopted sustainability and health as leading values in their business models. Lageweg (2017) predicts that health will be an additional accelerator in the creation of sustainable food chains. Healthy and fresh food lead to a smaller ecological footprint, promote shorter chains and strengthen the bond between producers and consumers. Conventional methods of food production cause all kinds of negative health effects, which can be substantial. This calls for considering the external effects in costing and pricing. New business models are appearing on the horizon based on collaborative actions by actors in the healthcare and food production sectors.

Climate change and human rights

Climate change induced by human activity is translated into broad programmes to reduce emissions of greenhouse gases (especially CO_2 and methane) by more efficient processes and using non-fossil renewable energy. It is important to realize that climate change relates to the previously mentioned normative issues that companies cannot ignore. Climate change affects people's livelihoods and how they experience human rights. Just think of the right to health, to water and sanitation, shelter, food, clean living conditions and opportunities for personal development. These are all recognized human rights that are under threat. For example, different parts of the world are undergoing a reduction in the availability of drinking water. Also, sea water levels rise while irregular weather affects harvests and food security. People who are affected the most by climate change are especially those who belong to the poorer segments of society, that is, women and children. Climate change has been caused primarily by the West, while most of its impact is felt in the developing countries. Therefore, the rich countries of the world, including the companies that originate or reside there, have a special responsibility to give support to the victims. Damage to human health caused by fossil fuels and climate change, in all regions of the world, is substantial. One may think of shorter lifespans, asthma, chronic lung diseases and the related reductions in the standard of living (de Graaf, 2015).

The sharing economy

The development of new business models that promote sustainability often comply with existing trends. One of these trends is the sharing economy, also called the peer-to-peer economy (de Waard, 2015; Davidson *et al.*, 2016).[2] Although the sharing economy is still a tiny part of the entire economy, it could grow substantially in the coming years. Well-known examples include Airbnb and Uber. However, it is not just rooms or cars that belong to the sharing economy; through Internet platforms, individuals share many assets with others. Consumers are active in both the demand and supply sides. They often become so-called 'prosumers', that is, consumers who have a clear influence on how the products they wish to buy are designed, produced and delivered. There are also new intermediaries that structure the market. By closely evaluating transactions and by naming and shaming through social media, high quality can be identified while poor quality and bad behaviour in the market can be controlled. Micro-insurance policies and guarantee funds can help contain risks. Moreover, governments can introduce regulations to reduce forms of nuisance and unwanted competition. A better use of existing capacities and goods is, in principle, in line with sustainability (Brachya and Collins, 2016). This should be viewed in relation to the lifespan of products and the effects of circular economy procedures.

If products were to last longer and be repaired more easily, households would have to spend less of their income on replacing everyday consumer goods that quickly become obsolete. The job creation potential in repair activities is huge. One difficulty faced by repair services today is the absence of repair manuals and spare parts associated with the product. However, there are regulations for the automotive sector which demand that manufacturers make repair information and spare parts available for all service providers. Within the EU, a solution could be to extend these regulations to household goods covered by the existing EU Eco-design Directive. Re-manufacturing a product goes one step further than repair. By disassembling a product and returning it to an 'as-good-as-new' state, it can be sold again on the market as a new product, thereby extending the lifetime of most of its parts. Re-manufacturing is currently working in business-to-business markets but it can, and should be, scaled up to consumer products (de Waard, 2015).

Various approaches to the circular economy run parallel with a growing knowledge and ability of recycling companies. One of these, Cradle to Cradle (C2C), has gained a great deal of attention in recent years (for more insight in this approach, see McDonough and Braungart, 2013). Although it has brought significant new insights and results, C2C is not a universal solution to resource-intensive consumption. In particular, the lifespan of products is a major factor in the eventual energy and material performance of products. The design process plays a crucial part in the eventual effect of circular modes of production. There should be a balance between technical lifespan and the time that a certain design style is appreciated by the public. The energy and material performance of products with a long lifespan is potentially much greater than that of C2C-based products with a short lifespan. Therefore, combining recycling with longevity will have a substantial positive effect on reduced material and energy consumption (van Dijk and Tubbing, 2013).

The turnaround of raw materials

The circular economy is rapidly becoming more popular. It is an economic and industrial system that takes its point of departure from the reusability of products and the resilience of natural resources. Value creation in each link of the system is aimed for, while minimizing the destruction of value in the entire system. Raw materials go through the following four turn-offs: disposal, incineration, recycling and new products based on 100% recycling. By looking at sectors in this way, you may discover where a sector's opportunities and threats lie for such a trajectory. The ultimate goal is new products based on 100% recycling. These new products to be produced require entrepreneurs, a proactive role for consumers and an integral provision of products and services. Here, the entire supply chain organizes itself around knowledge generation, design, financing, the winning of raw materials, production and sales (Büch, 2013).

The first steps towards the circular economy involve new products made of waste materials. In this way, waste is turned into new products. If this development takes its course and ends up in a mature market of demand and supply, all other turn-offs of roundabout can be omitted. This is an evolutionary process. It implies a shift in roles for the government, commissioners, contractors and consumers. In a circular economy, a central place should be given to how to get the markets up and running. Rules should be subservient to that central issue; it is no exception when existing laws and rules stand in the way of circular breakthroughs.

Innovation as an engine of sustainability

The canon of innovation

In many ways, sustainability innovations are part of the wider family of innovations. Therefore, understanding innovation at large is also insightful when it comes to sustainability innovation. Sustainability innovations can briefly be defined as innovations that are intentioned to have certain effects that promote sustainability.

In The Netherlands, a canon of innovation has been introduced that summarizes the outcomes of research on innovation over the last few decades (Verspagen *et al.*, 2013). Again, these insights are also relevant to sustainable innovation, especially when undertaken by profit-seeking enterprises. Table 2.2 summarizes the canon.

The electric car, which is expected to become the omnipresent replacement of the fossil fuel-based car that is so dominant today, is an interesting case of a radical innovation that requires system changes in the wider society (Wesseling *et al.*, 2014). See Box 2.1.

TABLE 2.2 Canon of innovation (Verspagen *et al.*, 2013)

1. **Innovation is no linear process**. There may different speeds along the way. The different steps are intertwined involving various actors. There may also be dead ends because knowledge and perspectives are imperfect.

2. **Radical and incremental innovations are cumulative and interdependent.** Radical innovation opens a variety of opportunities for incremental innovation.

3. **Equilibrium thinking (Walras) is less attractive than dynamic thinking (Schumpeter)**. Path-dependency in technological development, lock-ins by sunk costs and information asymmetry may reinforce competitive positions. Monopoly rents make the urge to innovate stronger.

4. **The patent system has many facets**. Patents can both promote and thwart innovation. So, a patent is a trade-off; the question is – how strong a patent should be in the interest of public welfare.

5. **Innovation is more than technological development alone**. It is embedded in the wider economic and social system. There is co-evolution of technology and societal institutions. At the macro level, we recognize this in innovation history. Also, at the micro level we see the interaction between technology and organization at work.

6. **Innovation activities are concentrated in the geographical space**. The world is a village, but the exchange of knowledge over long distances remains difficult. Many innovations depend on implicit, intangible and poorly documented knowledge (experience). Such knowledge is hard to exchange without personal contacts. That's why innovators with similar interests cluster in the same areas.

7. **The intertwined nature of innovation and the absorption of knowledge justify fundamental research.** If the country wishes to absorb knowledge produced elsewhere, it is also important to be engaged in fundamental research. Because of that, the distinction between innovation and absorption becomes less relevant.

8. **Product innovations conquer the market according to a fixed pattern.** New technology usually develops in three phases of a product's life cycle. Competition leads to a dominant design(s). Later these may cause lock-ins and loss of positions.

9. **Market structure and company size influence innovation efforts**. Companies play a major role in innovation, R&D being one of ways it can be realized. R&D volumes depend on the market structure and company size. In this type of analysis, the causality is a problem: market power can promote innovation; however, innovation can also be seen as an attempt to acquire market power.

10. **We cannot leave innovation to the market**. Innovation has positive externalities. As per the classical welfare theory, this will lead to market failure. Also, in ways other than through externalities, markets fail. In many cases there are no or insufficient institutions that shape the innovation process. An effective innovation policy involves major interferences in the market, and is not just following market forces. Learning and innovation are important, for policy makers, companies and knowledge institutes.

BOX 2.1 ELECTRIC CARS

Radical innovation undermines the competencies underlying a profitable competitive position of established companies. Considering the investments in electric cars by the 15 largest automobile manufacturers, it can be demonstrated that it was the economically less successful companies that focused on radical innovations.

One of the core objectives of Europe 2020 is stimulation of a sustainable energy. The electric car does not emit harmful gases and stimulates working towards a sustainable transport system. It is a radical innovation, as the established car manufacturers cannot use their competences around the combustion engine.

The innovation literature shows that the established manufacturers are not fully motivated to go for radical innovation as that would make their technology-specific knowledge superfluous and diminish their competitive advantage.

However, this picture should be refined. Conventional producers whose profitability was low were motivated to opt for radical innovation to a much greater extent than conventional producers whose profitability was high. This distinction between different established manufacturers is relevant to policy makers.

Wesseling *et al.*, 2014

Innovation is also undertaken by households for internal use, by public organizations (e.g. local governments, agencies) and by companies that have the primary goal of serving societal objectives (social innovation). Often, such innovation may be, technically speaking, simple improvement, but can engender substantial improvements in processes, products and services.

For a coherent discussion of innovation and sustainability, it is important to understand that all innovations may have both positive and negative consequences for people and their planet. Of course, there are many innovations that are not intended to serve sustainability objectives in the first place. Then, at least, sustainability features as a matter of avoiding or minimizing environmental and social harm or avoiding producing products that are not worth the sacrifice of the required resources. In other words, all innovations should be responsible innovation.

Responsible innovation

Responsible innovation has always been an important theme of research and innovation. Since the second half of the 20th century, science and innovation have become increasingly intertwined and formalized. Then, more than ever, it was realized that technology produces both benefit and harm; as a result, the debates about responsibility have broadened. Science and technology studies (STS) suggest that science and technology are not only technically but also socially and politically constituted. Paradoxically, science and technology can add to our sense of uncertainty and ignorance. Unforeseen impacts that are potentially harmful and transformative are likely to occur (Irwin, 2006). Emerging technologies often fall into an institutional void, that is, there are few agreed structures or rules that govern them (Hajer, 2003). This development has encouraged more decentralized and open-ended forms of governance involving a forward-looking view of responsibility and new forms of public dialogue.

Responsible innovation can be defined as taking care of the future through collective stewardship of science and innovation in the present (Stilgoe, *et al.* 2013). Table 2.3 presents a number of questions about responsible innovations that have emerged as important within public debates about new areas of science and technology.

TABLE 2.3 Questions on responsible innovations

Product questions	Process questions	Purpose questions
How will the risks and benefits be distributed?	How should standards be drawn up and applied?	Why are researchers doing it?
What other impacts can we anticipate?	How should risks and benefits be defined and measured?	Are these motivations transparent and in the public interest?
		Who will benefit?
		What are they going to gain?
		What are the alternatives?
How might these change in the future?	Who is in control?	
	Who is taking part?	
What don't we know about?	Who will take responsibility if things go wrong?	
What might we never know about?	How do we know we are right?	

Stilgoe *et al.*, 2013

The questions posed in Table 2.3 represent aspects of societal concern and interest in research and innovation. Responsible innovation can be seen as a way of including deliberation on these questions within the innovation process. Stilgoe *et al.* (2013) present a framework of four dimensions of responsible innovation for raising, discussing and responding to such questions. These dimensions are: anticipation, reflexivity, inclusion and responsiveness, and are briefly discussed below based on Stilgoe *et al.* (2013).

Anticipation

The detrimental implications of new technologies are often unforeseen, and risk-based estimates of harm have often failed to provide early warnings of future effects. Anticipation prompts researchers and others to ask 'what if' questions to consider what is known, what is likely, what is plausible and what is possible. It involves systematic thinking aimed at increasing resilience, while revealing new opportunities for innovation.

Reflexivity

This concept means holding a mirror up to one's own activities, commitments and assumptions, being aware of the limits of knowledge and the differences in perspective that people may have. Unlike the private, professional self-critique that scientists are used to, reflexivity is a public matter. Reflexivity demands openness and leadership within cultures of science and innovation.

Inclusion

The authority of expert, top-down policy making has been eroded. Therefore, there have been new voices in the governance of science and innovation, including voices from the wider public. Efforts have been made to develop criteria aimed at assessing the quality of this dialogue. Moreover, observed bottom-up changes within innovation processes may lead to greater inclusion. User-driven, open, open source and networked innovation seem to promote the inclusion of new voices in discussions of both the ends and means of innovation, but it is not clear whether that leads to a fundamental change.

Responsiveness

Responsible innovation requires the capability and willingness to change direction to respond to stakeholder and public values and changing circumstances. There are several mechanisms by which innovation can respond to improved anticipation, reflexivity and inclusion. In some cases, application of the precautionary principle, a moratorium or a code of conduct will be sufficient. Current approaches to technology assessment and foresight may be widened to create an improved responsiveness. Also, particular ethical values may be embedded in technology. Diversity is an important feature of responsive innovation systems. Existing policies, procedures and governance mechanisms in funding research and shaping technology may create types of unresponsiveness that choke creative ideas and innovation.

Various available mechanisms (such as public dialogue, research integrity, risk management and codes of conduct) can offer an overarching, coherent and credible governance approach to science and innovation only if they are aligned with one another. The integration of the previously discussed four dimensions can function as a general framework; however, to be effective, it needs to be adjusted to the specific circumstances under which decisions on science and innovation are to be made.

Social innovation

Social innovations are innovations developed and launched by entrepreneurs whose intention it is to solve certain societal problems. Making profits is a driving force, although creating a vital business that produces the new product or service may be part of the endeavour. Social entrepreneurs (those who are behind a social innovation) go through the same processes of entrepreneurship and innovation as profit-seeking entrepreneurs do but target their efforts in a different, socially valuable direction. Major social innovations include the kindergarten, the cooperative movement, first aid and the Fair Trade movement (Camilleri, 2017).

Often, social innovation is driven by individuals who have a passion for change with a notable and sustainable outcome. Part of this is user-led innovation; for instance, sufferers from a particular disease develop special electronic instruments by which they can ease inconveniences or contribute to well-being. In developing countries social innovation may take the form of, for example, low-cost irrigation technologies for subsistence farmers to survive dry seasons; the recycling of clothes and fabric to provide cheap clothing; forms of cheap healthcare; or micro-power plants to bring light and energy into the lives of rural populations (see www. ashika.org). There are also many forms of public sector innovations; most of the time these concern incremental improvements, for instance in communications (e.g. e-government), public transport, education and care (Camilleri, 2017).

However, social innovation has also become a notable component of large companies within the framework of CSR policies and earning acceptance by local communities (a 'licence to operate'). Moreover, (large) companies may be engaged in social innovation as a way of creating value to society which is greatly appreciated by their staff (in order to have better retention and create greater involvement). Social innovation can take place as a way of experimentation in providing useful new products to groups of people in need (e.g. healthcare for the rural poor or humanitarian emergencies such as floods or droughts). Often, such experimentation propels learning processes which later lead to up-scaling and commercial success (Camilleri, 2017).

Sustainability innovations

Responsible innovation and social innovation already involve innovations that promote sustainability. Sustainability is a major driver of innovation. For instance, this translates to pressures on organizations to change their products and processes to reduce CO_2 emissions and energy use. There are many innovation opportunities evoked by the prospective sustainable economy. Early efforts to address sustainability issues often centred on complying with new environmental laws as well as easy-to-do activities by which companies tried to improve their image. Nowadays stricter environmental laws are still a major driver of innovations. However, rather than doing what is minimally required, leading companies recognize the need for dealing with resource instability and scarcity, energy security and systemic efficiencies across their supply chains, and to develop their innovations accordingly.

Table 2.4 gives examples of sustainability-led innovation based on four different innovation targets (Bessant and Tidd, 2015).

Table 2.4 indicates that in addition to product/service and process innovation, changing of position and introducing new business models (paradigm innovation) can play a prominent role in sustainability-led innovation.

Sustainability innovation can be highlighted by three dimensions that 'underpin a change in the overall approach from treating the symptoms of a problem to eventually working with the system in which the problem originates' (Camilleri, 2017). There are all kinds of possibilities to mitigate environmental and social problems, starting with somewhat defensive policies entailing compliance with existing environmental laws and improving existing processes leading to greater efficiency (saving raw materials and energy, producing less waste and better controlled waste management). The second phase is an organizational transformation leading to new products and/or services, new business models which include forms of shared value. Eventually, sustainability inspires the building of new

TABLE 2.4 Examples of sustainability-led innovation

Innovation target	Examples
Product/service innovation	Green products, design for greener production and recycling, service models replacing consumption ownership models.
Process innovation	Improved and novel manufacturing processes, lean systems within the organization and across the supply chain, green logistics.
Position innovation	Rebranding the organization as green, meeting needs of underserved communities (e.g. bottom of pyramid).
Paradigm innovation – changing business models	System-level change, multi-organization innovation, servitization (moving from manufacturing to service emphasis).

Bessant and Tidd, 2015

systems that go beyond the individual firm because the game-changing innovation required by a fully fledged transformation towards sustainability does not strive in isolation; instead, they require a variety of other parties that contribute essential knowledge, capacities and facilities or provide additional institutions. Also, the concepts of sustainable cities and sustainable communities are of relevance here – that is, local non-profit organizations, community-based organizations and responsible civilians can be essential partners in realizing sustainability through innovation (Wesseling *et al.*, 2014).

The previous sections have shown that sustainability innovation can be provided by different types of entrepreneurship. It might be start-ups or incumbents, profit-seeking or social endeavours, local community-based or global initiatives. Innovation networks in national or regional settings may be a major force to take sustainability to the next level.

Conclusion

This chapter has discussed various ways in which sustainable value can be created. It is a mixture of private and public efforts to turn an unsustainable economy into a sustainable one. The mixture refers to combinations of values, capacities, opportunities and threats. This variety invites a pallet of institutional forms of organization, profit-based, not-for-profit, private, community-based, public, large, small, various kinds of innovation systems and networks. They are pathways to overcome the business-as-usual approaches that still dominate today's economies. Innovation is a major transformation route, but will be effective only if and when society wants it and sets adequate priorities.

Not least because of the limited results that recent efforts to implement sustainability have yielded, there is a tendency to place a greater emphasis on the aspirational and actual societal benefits of human activities. A substantial part of the societal benefits refers to what is implied in sustainability. For companies, this means that their business cases cannot begin with company-based benefits but are ever more expected to take societal benefits as a starting point for business development. From there, an integration of business interests and wider societal interests should take place.

The question arises as to whether today's would-be business managers are adequately educated to take that 'reversed sequence' into account. Contemporary dominant business paradigms strongly derive from the neoliberal narrative.

Neoliberalism entails a particular interpretation of the core values of freedom and democracy; it demands that businesses and economies can thrive only under conditions of free markets and private property, free and rational individuals, free trade, a small government and a reduced social welfare (Waddock, 2016).

In the past, neoliberalism seemed to be successful in unleashing economic growth around many parts of the world, but now we see its negative side: it led to a considerable degree of inequality and instability and greatly contributed to climate change and an overuse of natural resources. That is why human civilization needs a system change today. Part of that change is exploding neoliberal myths and adopting a new narrative embracing values such as well-being, dignity, diversity and harmony with nature (Waddock, 2016).

Coming back to business education, one should ask whether present educational programmes are promoting a mindset that makes it possible for future managers to lead the transition. By all means, this should not be a move away from liberty and democracy but a renewed narrative of inclusive prosperity in which markets and business are not such an overriding factor.

In this respect, Laszio *et al.* (2017) see business schools torn between two paradigms with a resulting struggle about the nature and value of both teaching and research. Indeed, they note that today's dominant neoliberal paradigm pervades the vast majority of business schools with its narrative of profit maximization, free markets and limited government. The authors state that because the dominant narrative's assumptions are so pervasive, most of us accept them without questioning them. They present as a counterpoise to neoliberalism the emerging 'economy in service of life', which is an alternative that attempts to consider the complex realities of the contemporary world, involving the sustainability imperative as discussed in this chapter; this alternative entails flourishing by no longer merely attempting to do less harm, but instead to set at the core the goal of sustaining the potential for human and other life to flourish on this planet (Ehrenfeld & Hoffman, 2013).

Since the turn of the century, business education has been confronted with a calling for more transformational initiatives. Especially, millennials have appeared to be interested in (working for) a responsible business, while educational programmes in business ethics have seen the largest growth compared to other subjects. However, many business schools seem to delay while accommodating the new emergent paradigm may be the greatest moral challenge of our time. Taking action will involve a rethinking of the traditional siloed courses. However, the transition towards a sustainable society is not a matter of choosing between opposing paradigms but a matter of a fully integrated curriculum that transcends the two paradigms. Only then will our students be prepared for leadership roles and the complexities of the world in which they live (Laszio *et al.*, 2017). For instance, such roles have to be played by CEOs, CFOs and controllers. They can be the intermediaries that bring together the representatives of the various professional functions in a company under the common banner of sustainability (Wolters, 2013, chapter 1).

Corporate sustainability requires an ethics that is prospective (Wempe, 2016). Prospective ethics is about the role that an individual, a company or any other organization chooses to play in society. For our schools and educators this translates into the question: do we prepare our students for a sustainable future?

Notes

1   This does not imply we can be sure that the sustainability steps that are being taken will be sufficient to prevent drastic climate change and serious loss of biodiversity. Here we touch on the micro–macro issue.

2   According to the *Oxford English Dictionary*, the sharing economy is defined as 'an economic system in which assets or services are shared between private individuals, either for free or for a fee, typically by means of the internet'. Traditionally characterized as a peer-to-peer resource network, this model is most likely to be used when the price of an asset is high and the asset is underutilized or is operating at idle capacity.

References

Ayers, R.U. (2014) *The Bubble Economy. Is sustainable growth possible?* Cambridge, MA: The MIT Press.

Bessant, J. and Tidd, J. (2015) *Innovation and Entrepreneurship*. Third edition. Chichester, UK: John Wiley.

Brachya, V. and Collins, L. (2016) *The Sharing Economy and Sustainability*. Jerusalem: The Jerusalem Institute for Israel Studies.

Büch, R. (2013) Wegwijzers op de grondstoffenrotonde, *ESB*, 19(8), pp. 25–7.

Camilleri, M.A. (2017) Corporate sustainability and responsibility: creating value for business, society and the environment, *Asian Journal of Sustainability and Social Responsibility*, 2, 59–74.

Carroll, A.B. (1979) A three-dimensional conceptual model of corporate performance, *The Academy of Management Review*, 4(4), pp. 497–505.

Carson, R. (1962) *Silent Spring*. Boston, MA: Houghton Mifflin.

Cramer, J. (2013) Duurzame stad heeft de toekomst, *Tijdschrift Milieu*, 19(1), pp. 22–3.

Crane, A., Palazzo, G., Spence, L.J. and Matten, D. (2014) Contesting the value of 'creating shared value', *California Management Review*, 56(2), pp. 130–53.

Dagevos, J, Smeets, R, Mulder, R. and Janssen, J. (2017) Stad behoeft krachtige sociale agenda, *Tijdschrift Milieu*, 23(1), pp. 12–15.

Davidson, N.M. and Infranca, J.J. (2016) The sharing economy as an urban phenomenon, *Yale Law & Polity Review*, 34(2), Article 1, pp. 215–79.

Dijk, G. van and Tubbing, A. (2013) Afvalhout ontbeert, *Tijdschrift Milieu*, 19(8), pp. 18–19.

Ehrenfeld, J. and Hoffman, A. (2013) *Flourishing: A Frank Conversation about Sustainability*. Palo Alto, CA: Stanford University Press.

Galant, A. and Cadez, S. (2017) Corporate social responsibility and financial performance relationship: a review of measurement approaches. *Economic Research – Ekonomska Istrazivanja*, 30(1), pp. 676–93.

Graaf, J. de. (2015) Mensenrechten komen zwaar onder druk, *Tijdschrift Milieu*, 21(7), pp. 8, 9.

Hajer, M. (2003) Policy without polity? Policy analysis and the institutional void, *Policy Sciences*, 36, pp. 175–95.

Henderson R. (2015) Making the business case for environmental sustainability. In: Henderson, R. et al. (2015). *Leading Sustainable Change. An Organizational Perspective* (Ch. 2). Oxford: Oxford University Press.

Irwin, A. (2006) The politics of talk: coming to terms with the 'new' scientific governance, *Social Study of Science*, 36, pp. 299–330.

Laasch, O. and Conaway, R.N. (2015) *Principles of Responsible Management. Glocal Sustainability, Responsibility, and Ethics*. Stanford, CT: Cengage Learning.

Lageweg, W. (2017) Gezondheid als driver voor duurzaamheid, *TGTHR, 22 November 2017*. Available at https://tgthr.nl/gezondheid-als-driver-duurzaamheid/ (accessed 13 December 2017).

Laszio, C., Sroufe, R. and Waddock, S. (2017) Torn between two paradigms: A struggle for the soul of business schools, *AI Practioner*, 19(2), pp. 108–17.

Margolis, J.D. and Walsh, J.P. (2003) Misery loves companies: Rethinking social initiatives by business, *Administrative Science Quarterly*, 48(2), pp. 268–305.

McDonough, W. and Braungart, M. (2013) *The Upcycle. Beyond Sustainability – Designing for Abundance*. New York: North Point Press.

Porter, M.E. and Kramer, M.R. (2006) Strategy & society. The link between competitive advantage and CSR. *Harvard Business Review*, December 2006.

Porter, M.E. and Stern, C. (2017) *Social Progress Index 2017*. Washington, DC: Social Progress Imperative.

Schumpeter, J.A. (1934) *Theory of Economic Development*. Cambridge, MA: Harvard University.

Staps, S., Wietmarschen, H. van and Lubbe, J. van der, (2017) Positieve benadering gezondheid werpt vruchten af, *Tijdschrift Milieu*, 23(1), pp. 28–9.

Stilgoe, J., Owen, R. and Macnaghten, P. (2013) Developing a framework for responsible innovation', *Research Policy*, 42, pp. 1568–80.

The Economist Intelligent Unit Ltd (2017) *The Road to Action. Financial regulation addressing climate change*.

Verspagen, B., Kleinknecht, A. and Frenken, K. (2013) Canon deel 6: Innovatie, *ESB*, *98* (4656), pp. 184–187.

Waard, M. de (2015) De economie van Airbnb, *ESB*, 100 (4722), pp. 696–7.

Waddock, S. (2016) Foundational memes for a new narrative about the role of business in society, *Humanist Management Journal*, 1, pp. 91–105.

Wempe J. (2016) The serving university: A matter of prospective ethics. In: M. Flikkema *et al.* (2016). *Sense of Serving. Reconsidering the Role of Universities Now*. Amsterdam: VU University Press.

Wesseling J., Niesten, E., Faber, J. and Hekkert, M. (2014) Prikkels en kansen voor duurzame innovatie, *ESB*, 99 (4676), pp. 10–12.

Wolters, T. (2013) *Sustainable Value Creation as a Challenge to Managers and Controllers*. Apeldoorn, Netherlands: Wittenborg University Press.

3

SUSTAINABLE STRATEGIES AND ACCOUNTING NEEDS

Teun Wolters and Jan Jaap Bouma

Sustainability between hope and fear

Sustainability is about a future that requires action today. We find ourselves between hope and fear, between activism and the realization that we do not do enough. Research institutes that follow trends and measure progress in sustainable development exhort governments to swing into action. For instance, drastic interventions in agriculture and farming are pleaded. Climate change, biodiversity and the wanted, but mostly slowly evolving, circular economy is a matter of great concern; we're lagging behind. In many places, in the world of governments, there are many policies in place that promote sustainability and there are results to be reported. However, each time these results appear to be insufficient to guarantee a sustainable future.

In the business community, we see similar trends. Many companies have introduced policies and practices that relate to sustainability. There is progression in many areas; sometimes leading companies attract attention because of their progressive strategies. Nonetheless, many companies still take a passive stand on sustainability. They often intend to obey the law at least and try to avoid courses of action that could harm their reputation. However, even such middle-of-the road behaviour cannot be taken for granted. The trickeries discovered undertaken by Volkswagen and, as it appeared, other car makers (to hide the true levels of harmful content of the exhaust gases) are spectacular, but at the same time just one of many manifestations of short-termism and lack of social responsibility that characterizes much corporate activity. Also, here, we find ourselves in a position between hope and fear. True, such problems could later speed up the introduction of cleaner technologies because of increased public awareness and political resolve.

In sustainability research, we see parallel developments: Repeatedly, new concepts or tools are launched because previous means, even if implemented to the full, were deemed insufficient to reach sustainability. In the 1990s, for instance, the implementation of environmental management systems was encouraged. However, it did not take long before there were mitigating voices: such management systems did not require ambitious goals while it was even possible to 'go through the motions' to please the market without many real implications for the running of the business. Here, we come across a 'system fatigue' that may happen in all sorts of situations.

In recent times, there has been a plea to broaden corporate strategy and integrate various aspects and functions in it. For instance, the balanced scorecard (BSC) was developed in the early 1990s as a new approach to performance measurement due to problems of short-termism and an orientation towards the past in management accounting (Figge et al., 2002). Quality and process improvement programs existed well before the introduction of the BSC. In the 1970s, Japanese companies demonstrated the strength of their total quality management (TQM) approaches, which built on earlier innovations by Deming and others. Western versions of Japanese TQM appeared in the 1980s, including lean management, just-in-time and six sigma (Kaplan and Norton, 2008). Quality models focus on identifying processes that fall short of best practices. This focus, however, occurred independently of strategic priority setting whereas the BSC provides explicit causal links from quality and process improvements to successful outcomes for customers and shareholders. The highlighted cause-and-effect relationships indicate the process improvements that are most critical for successful strategy execution. Meanwhile, in research at least, the BSC has become a prominent managerial tool in directing a company to sustainability and continuity.

The conventional BSC, as a tool of performance management, does not explicitly distinguish between different stakeholder interests or visibly integrate sustainability into a company's strategic goals. However, the sustainable BSC (SBSC) has become a successful exponent of taking corporate sustainability into account. Under that banner, corporate sustainability can benefit from the connections between strategy and operations inherent in the BSC. The interpretation of the term 'balanced' is extended to include objectives of corporate sustainability (Möller and Schaltegger, 2005). That means the SBSC helps overcome the shortcomings of conventional approaches to environmental and social management systems by integrating the three pillars of sustainability (economic, social and ecological dimensions) into a single and overarching strategic management tool (Figge et al., 2002).

This chapter focuses on strategic decisions that further integrate sustainability into a company. This is seen as a necessary development. Accounting is to deliver the information that the decision makers need in order to underpin their decisions. Where decisions on sustainability involve a wider community, the availability of accounting information that is accepted by that community is of great importance. This chapter pinpoints various developments in accounting that are needed or desirable in this context. Most of all, it aims at indicating where accounting needs are to follow the strategies that are capable of bringing sustainability to realization.

Continuity and wider frameworks

Strategic decision making aims at the continuity of the firm by creating value for its stakeholders. From the perspective of nature and its resources, strategic decision making of individual firms can easily be too narrow. The governance of natural resources calls for various transitions, such as those to non-fossil energy and sustainable water management. These transitions require innovative business models involving the adoption of new technologies and the use of renewable resources. Such business models should be based on a willingness to take social responsibility and be open to society, which, by the way, has clear consequences for a company's external reporting. Business models will be further discussed later in this chapter.

Capital budgeting is that part of management accounting that can be used to allocate capital resources to technological innovations, inclusive of innovations that reduce environmental impacts and the need for water, and increase water efficiency, recycling, reusage and recharging of local water resources. Performance indictors may include a broad set of impacts on a company's natural environment. Cost accounting may contribute to reporting on a company's levels of efficiency and eco-efficiency.

Kaplan and Norton (2008, p. 159) stress that a strategy can only be implemented if it is linked to excellent operational and governance processes. The operational side of strategy is represented by the BSC or SBSC. However, the question may arise as to whether a company's strategy is sufficiently visionary to stand the test of time and contribute to a sustainable future. It seems that there is a need for wider frameworks that clarify what sustainability implies and make it possible to position companies in terms of 'distance to target'. In these wider frameworks, instruments such as the SBSC and Business Models for Sustainability can have a place (Möller and Schaltegger, 2005).

When considering such wider frameworks, it seems worthwhile to pay attention to the concept of corporate governance. Moreover, the notion of a stepwise development towards integrating sustainability in the company seems to be a helpful approach to understanding how a company is moving towards sustainability in the fullest sense of the word.

Corporate governance and sustainability

Corporate governance is a matter of administering a company within the confines of the law and acceptable ethical principles. Historically, the interests of shareholders received top priority. Over the last few decades, corporate governance has been reconceptualized to include various stakeholders, that is, individuals and organizations capable of influencing the company and vice versa (Pettigrew, 2009; Freeman *et al.*, 2010).

Therefore, corporate governance is a part of a company's business and society relations. It falls within the range of issues concerning corporate power, legitimacy and responsibility (Pruijm, 2010).

Corporate governance can be evaluated from different angles. Pruijm (2010) distinguishes:

1   A business-administrative point of view: there is a need for proper management based on well-thought-of strategies, motivating leadership and effective management systems.
2   Economically speaking, the company is to create economic value that can justify its existence.
3   Corporate sustainability, implying that value creation must be sustainable in line with the three P's (people, planet and profit). Reputation may be at stake.
4   From a juridical point of view, a company board is responsible for what happens in a company; it represents the company as a legal identity.

In a sense, corporate governance coincides with effective management control, the CFO and the controllers playing a significant role in it. This role is not mere performance measurement but evolves into performance management.

According to Strikwerda (2009), today, in the conduct of business, performance management is to be embedded in the totality of a company's mission, values and vision. 'Mission' indicates what the owners and management imagine they want the company to be and what they see as their commercial and societal assignment. Values express a company's responsibility vis-à-vis various stakeholders. A company's vision is its concrete medium-range view of the market and its wider economic environment. Vision is fundamental to formulating strategic choices. In this way, measurable goals, both financial and non-financial, are embedded in a value system and placed within a certain time frame. In Strikwerda's view, via a business model or strategy map, the measurable goals are also underpinned in causal terms, because then they show how a company's capacities and views make themselves felt in operational processes and customer satisfaction. This approach allows a company to make clear to its external stakeholders what it does (and does not) stand for. Internally, it enables a company to explain what its mission and ranked values mean and how they relate to the achievements of its employees. The ranking of values (e.g. whose interests come first – shareholders, customers or employees) is considered of importance as it makes it possible to check whether the management sets its priorities right (see Table 3.1). It is striking that Strikwerda explicitly considers the personal values of managers and possibly other decision makers. This assumes that the influence of organizational goals can be either supported or thwarted by personal values and ambitions. Ideally, the entire structure leads to satisfied stakeholders, satisfied clients, efficient and effective processes and motivated co-workers. However, personal values are difficult to capture and map. These values may change over time and are subject to influences by other stakeholders of the firm. Values of stakeholders may conflict with each other in both the short and long term.

In each of the elements of Table 3.1 (the steps from mission to strategic results), sustainability can play a role. When going into the details of an individual company, the role of sustainability can be made visible.

TABLE 3.1 The steps of strategy and performance management embedded in non-financial values and norms

1	2	3	4	5	6	7
Mission	Ranking of values	Vision	Strategy	Balanced Scorecard	Operational goals and initiatives	Personal goals

8.0	8.1	8.2	8.3
Strategic results:	Satisfied stakeholders (no clients)	Satisfied clients	Motivated co-workers

Based on Strikwerda (2009)

According to Strikwerda (2009), this approach also reduces the risk that internal budgeting processes and the definition of tasks for each department are manipulated by managers for the benefit of their personal agendas, which may be at odds with the company's interests. Parameters such as a business unit's profit, shareholder value or ROI are susceptible to inexpedient decisions when separated from corporate values and (possible) social and ecological effects.

Potential employees can decide for themselves whether a company's values match with their own personal values. When assessing shares, shareholders can consider a company's values and how they are ranked. The same applies to customers and suppliers. Similarly, public opinion can make itself known by expressing judgements on these values. The results of all these considerations inform the company about its actual contribution to the development of prosperity (in a broad sense of the word).

Responsible management

The extension of corporate governance towards a broad group of stakeholders has led to an even further broadening of corporate concerns and responsibilities. This movement has inspired the areas of stakeholder management, business and society relations and, not least, Corporate Social Responsibility (CSR).

ISO 26000 provides clarity on both the content and the process side of CSR. It distinguishes seven general principles of social responsibility that are applied to seven core subjects (which are worked out into CSR issues). The seven general principles are: accountability, transparency, ethical behaviour, respect for stakeholder interests, respect for the rule of law, respect for international norms of behaviour and respect for human rights. Moreover, ISO 26000 holds more specific principles, such as environmental principles. There is no need to state that such a systematic overview of CSR activities can serve the purpose of sustainability accounting. In the area of external reporting, the Global Reporting Initiative (GRI) provides highly usable models to cover the broad area of corporate sustainability and responsibility.

The broadening of frameworks and fields of special attention seems to have culminated in what is called Responsible Management. In Responsible Management, the individual perspective has a prominent place (which is also reflected in Table 3.1 by distinguishing personal goals). Laasch and Conaway (2015) claim that the logical next evolutionary step is translating the organizational vision into managerial and operational achievement by a responsible manager, but without giving up the merits of the previous organizational emphases. These authors distinguish three domains:

- Sustainability (aims to create a neutral, or better, a positive triple-bottom line)
- Responsibility (aims to optimize stakeholder value)
- Ethics (aims to create moral excellence).

These domains are worked out based on distinguishing three levels (Table 3.2): the organizational level, the management level and the core concept level.

The above is just to give an impression of the Responsible Management approach (for the details, see Laasch and Conaway, 2015). It is a framework that covers a large variety of topics and issues relating to sustainability and CSR.

Part of the Responsible Management 'programme' is achieving stakeholder accountability by integrating sustainability, responsibility and ethics into the company's accounting and controlling systems. Along with this, it is necessary to develop and use indicators for social, environmental and ethical activity and performance as a basis for Responsible Management activity. There is a need for internal and external reporting on Responsible Business and its performance (Laasch and Conaway, 2015). 'Sustainability accounting and controlling' is defined as a subset of accounting and reporting that deals with activities, methods and systems to record, analyse and report: (1) environmentally and socially induced economic impacts; (2) ecological and social impacts of a company, production site, etc.; and (3) measurements of the interactions and links between social, environmental and economic issues constituting the three dimensions of sustainability.

The authors bring to the fore that in their view the dominant paradigms of financial management (concerning profit, growth, short-term, money as a

TABLE 3.2 Levels and domains of responsible management

Level	Domain		
	Sustainability	*Responsibility*	*Business ethics*
Organizational level	Business sustainability	Business responsibility	Business ethics
Management level	Sustainability management	Responsibility management	Ethics management
Core concept level	Triple bottom line	Stakeholders	Ethical issues

Laasch and Conaway, 2015

measuring rod, shareholders) stand in the way of creating a responsible business. Solutions to dismantle those paradigms might require dramatic changes and disruptive innovation. However, such solutions are not readily available, which justifies continuous research efforts to find or construct them (Laasch and Conaway, 2015, ch. 15).

The CFO was already mentioned when discussing the critical role of CFOs and controllers in ensuring management control in support of corporate governance. The modern CFO is also to bridge the gulf between those who articulate sustainability goals and conditions, on the one hand, and the financial specialists whose financial frameworks (and mental frames) do not (sufficiently) incorporate sustainability values, on the other. This connection shows how the functional and personal value systems relating to finance and financial management can play a crucial role in how a company develops. This CFO role is a crucial and at the same time vulnerable aspect of achieving a sustainable enterprise (see also Wolters, 2013, ch. 1). At some point, this intermediating role of the CFO should end with a unified approach to sustainability in the company, including the use of appropriate sustainability-based KPIs.

Similarly, controllers can play a significant part in the field of sustainability and its accounting. The following possible controller roles can be noted (Wolters, 2013, ch. 1):

1   Contributing to the formulation of a strategy that incorporates sustainability (such as calculation of costs, cost avoidance and budgets).
2   Giving investments in sustainability a fair chance (in spite of conservative and routinely handled criteria regarding timespan, hurdle rates and discount rates).
3   Highlighting the benefits of preventative measures by making sure that all risks that they reduce or avoid will count.
4   Accounting in support of the greening of operational production processes, such as material flow accounting and making visible the various quantitative aspects of sustainability.

Technology change, strategic and organizational adaptation

One of the issues that needs to be discussed is that, especially in industries where advanced technology is a prominent part of intense competition, the 'static' identification of competitive advantage could easily overlook an important strategic opportunity. Depending on the industry and market, strategy may be changing all the time because of external economic pressures, decisions on greening decisions and marketing ambitions. This may require striking a balance between, on the one hand, a structured approach to strategy based on existing competitive advantage, and experimenting (the non-static dimension) with entirely new technology and business models on the other hand (Lynch, 2015).

As competitive advantage is constantly evolving, strategies need to be proactive. Particularly in highly competitive, if not hyper-competitive, markets, innovations occur frequently. Hyper-competition means constantly escalating rivalry caused by rapid product innovation, shorter design and product life cycles, aggressive price- and competence-based competition and experimentation with new approaches to serving customer needs.

For companies to survive and be successful, dealing with hyper-competition requires market dominance. Operating under such circumstances implies that change initiatives like re-engineering, continuous improvement and employee empowerment are not sufficient to survive and thrive (Volman, 1996). Under hyper-competitive circumstances, the alternative to market domination, Volman argues, is organizational death, which may either happen quickly or, more frequently, slowly. The signs of organizational decay are often identifiable long before crises become apparent, considering signals such as a loss of market share, possession of the wrong set of competences, slowing growth patterns, a loss of employee morale, poor product and process development and the failure to recognize competitors' true capabilities. Only when change programmes are deep and fully integrated across the organization can an enterprise be truly transformed (Volmann, 1996). Part of the transformation is the ability to engage in early visionary thinking about new kinds of business models, because waiting for the moment of obvious business opportunity may be too late considering the speed that is needed. This forward-looking aspect of operating in hyper-competitive markets requires a rethink of the value of using tools such as the strategy-based BSC and the SBSC. There may be insufficient time to go through the entire process of strategy formation and developing KPIs. Where strategies becomes fluent, performance management may require a different approach.

This corresponds with a critical view on strategy as a planning process. According to this view, by its very nature, strategy making is uncomfortable and apprehensive; it is about making hard choices; it is about thinking through what it would take to achieve what you want and then assessing whether it is realistic enough to give it a try (Martin, 2014). Here, management accounting may be somewhat distanced from formal strategic frameworks; perhaps, it should be directly linked with the preceding values and objectives, on the basis of which waves of strategy can be monitored and evaluated. Of course, values and objectives in the area of sustainability can be part of the endeavour.

However, sustainability does not only play a significant role in highly competitive markets; other markets may develop in ways where market domination is not a *sine qua none*. It can be assumed that technological development and change take place in all industries and markets. However, there are sectors where changes occur more slowly and therefore make it possible to take time to research and develop new ways of doing business, in particular sustainable business. Sustainable innovations which relate to system change need to be embedded in wider societal adaptation and collaboration with other partners, such as governments, NGOs and community-based initiatives. Here, speed is not necessarily the greatest worry but could be instead a matter of operating in contexts which abound in complexity and dynamics. Complexity and dynamics go together with uncertainty. Dealing with uncertainty (also, in terms of timing) can be a puzzling reality for many companies.

There is room for considering the relationship between sustainability and external uncertainty. In many areas of sustainability, there are developments which make it difficult to come to grips with the costs and benefits of different solutions (Laasch and Conaway, ch. 14):

1 Many green technologies are still the subject of basic research and development, while for the more proven technology there is ongoing advanced production technology research.
2 Costs of energy are uncertain. In the longer term, costs of energy may rise while current costs are relatively low.
3 The economies of scale associated with existing green technology will reduce costs over time as companies engage in larger-scale production.
4 Stakeholder attitudes to sustainability, including those of customers, are still developing.

Hence, the benefits and costs of sustainability strategies are uncertain and must be considered, not only from a prescriptive but also from an emergent point of view. A prescriptive strategy is a strategy whose objective has been defined in advance and whose main elements have been developed before the strategy commences. After defining the objective, the process includes analysis of the business environment, the development of strategic options and the choice between these. The chosen strategy is then implemented (Lynch, 2015, ch.2). Mintzberg (1990) identified major difficulties with this prescriptive strategic process, such as less knowledge and predictive power than assumed and less agreement in the organization about the strategy than assumed. Separating strategy formulation and implementation as distinctive phases may be too simplistic in many complex strategic decisions. This may lead to the conclusion that a company should focus solely on emergent strategy or adopt a combination of both prescriptive and emergent strategy. Under emergent strategic management, a strategy has no final objective: its elements are developed in the course of time as the strategy proceeds (Lynch, 2015, ch. 2).

These dynamic and emergent elements of strategy seem to render less straightforward the relationship between strategy and implementation by the corporate organization and its partners than is often assumed. Here, the organization's strength comes into play. A concept that could enlighten this aspect will be discussed here – it is the concept of organizational agility (Dyer and Shafer, 1999), which is the ability to rapidly adapt one's organization to different external circumstances as a matter of strategic competence. Agility makes it possible to deal with sudden change and reduces an organization's dependence on strategic management processes because change can be relatively easily accommodated by the organization while maintaining the given objectives and the KPIs that directly relate to them.

The agile organization (and in fact its personnel) conforms to the following characteristics (Boxall and Purcell, 2011):

1   There is a high level of initiative (the organization is undertaking and enterprising).
2   There is a tendency to work together spontaneously.
3   There is an ability to renew the organization (to innovate through knowledge and mentality).
4   There is an ability to take up new activities rapidly and repeatedly.
5   The workers are goal-oriented (sustainability can be one of the goals, possibly differentiated into different sub-goals).
6   The workers are ready to work in teams.
7   The personnel accept that developments can at times be paradoxical.
8   There is a sense of operating in a business-like manner (costs and benefits have to match).

These characteristics could be worked out for a company operating in versatile markets and intending to integrate sustainability in its strategies, business models and organization. These characteristics could be defined as major KPIs and repeatedly measured, starting with a baseline measurement.

Business models

Not only do BSCs have a direct tie to strategy. This connection also refers to the phenomenon of business models. The starting point is the core business strategy of the company. Every part of the business model must be consistent with this core. Analysis of the existing business model can be the beginning of developing new insights into whether there are new market opportunities for the company. The details of a business model may make them long and complicated. Rather than leaving out significant aspects, it is preferable to consider the details but then, for the sake of communication and further decision making, a business model must later be simplified, either by aggregation or decomposition into elements that do not interact (Lynch, 2015, ch. 20).

Involving sustainability in a business model can be done if it is part of the core strategy. Then sustainability will be inevitably on the agenda. There is significant research done on the development of business models for sustainability (Schaltegger et al., 2015) and further research can be expected to emerge. For instance, Abdelkafi and Täuscher (2016) have developed a conceptual business model for sustainability (BMfS) that incorporates the natural environment as an essential element. This model shows the complexity involved and the difficulty to avoid that – the approach is interesting but limited in scope (limited to the natural environment); nonetheless, it is relatively complicated.

Sustainable business model innovation

Sustainability leaders know that many existing business models are predicated on the erroneous assumption that natural and social capital are in almost limitless supply while mispriced resources and other market failures make some models more attractive than they deserve to be. Therefore, there is an urgent need for fundamentally different approaches to value creation (SustainAbility, 2014). These new approaches and the business models based on them are already appearing. How companies engage in a transformation process by abandoning conventional modes of value creation and adopting a new business model for sustainability is not yet fully understood. Roome and Louche (2016), based on casework in the production of textiles and in the construction sector, suggest a process model that shows how a new business model develops. This can be seen as a further exploration of what new business models require and imply. This process model can be summarized as follows:

1   There comes a moment when a company realizes that it has to deal with a new, complex and changing economic, environmental and social context, which places demands on it that its traditional approach to business cannot fulfil.
2   This translates into a new vision and concepts for the company by learning and action, which is the basis for new capacities and knowledge.
3   Through a tied network of communities within the company, the new capacities and knowledge are embedded in the company and applied to a new business model and the abandonment of the old business model.
4   Further external communication and diffusion lead to broader networks in support of the new business model.

SustainAbility (2014) distinguishes 20 business models that produce more sustainable outcomes; these 20 models fall into 5 categories (see Table 3.3).

Business models and accounting

Three dimensions of strategic planning

The amount and particular content of the accounting information, in a broad sense of the term, needed to support the adoption and implementation of a strategy and a related business model is contingent on various factors. Strategic planning, including strategic marketing planning, can be divided into three dimensions (Piercy, 2017):

1   An analytical dimension (a series of techniques, procedures, systems and planning models)
2   A behavioural dimension (the nature and extent of the participation, motivation and commitment among members of the management team)

TABLE 3.3 Five categories of sustainable business model innovation

Category of sustain-able business models	Breakdown	Brief explanation
1 Environmental impact	Closed-loop production	Continuous recycling through the production system
	Physical to virtual	Replacing 'bricks and mortar' with virtual services
	Produce on demand	Producing only when consumer demand is clear and confirmed
	Re-materialization	Innovative ways to source materials from recovered waste, creating entirely new products
2 Social innovation	Buy one, give one	Selling a specific good and donating a similar good to those in need
	Cooperative ownership	Companies owned by members, taking wider community concerns into account
	Inclusive sourcing	Making the supply chain more inclusive. e.g. helping farmers providing the product
3 Base of the pyramid	Building a marketplace	Companies build new markets for their products in innovative and socially responsible ways, serving poor people
	Differential pricing	Companies charge more to those who can afford it to subsidize those who cannot
	Micro-finance	Providing small loans and sometimes or financial services (e.g. insurance) to low-income borrowers with no access to traditional banks
	Micro-franchise	In line with traditional franchising, but specifically focusing on creating opportunities for the poor to own and manage their own businesses
4 Diverse impact	Alternative marketplace	Invention of a new type of transaction to unleash untapped value

continued . . .

TABLE 3.3 Continued

Category of sustain-able business models	Breakdown	Brief explanation
	Behaviour change	Using a business model to stimulate behaviour change to reduce consumption, change purchasing patterns or modify daily habits
	Product as a service	Consumers pay for the service as product provides without the responsibility of repairing, replacing or disposing of it
	Shared resource	Enabling customers to access a product rather than own it and use it only as needed
5 Financing innovation	Crowdfunding	Enabling a (social) entrepreneur to tap the resources of his/her network to raise money in increments from a group of people
	Innovative product financing	Consumers lease or rent an item that they cannot afford or do not want to buy outright. Could go together with 'product as a service'
	Pay for success	Employing performance-based contracting, typically between providers of some form of social service and the government

Sustainability (2014)

3   An organizational dimension (information flows, structures, processes, management style and culture).

Companies differ in all of these dimensions and where their strengths and weaknesses lie. In terms of integrating sustainability in the company, the three dimensions are of paramount importance: they determine the openness, readiness and capacity to make corporate sustainability real.

In terms of behaviour and information, managers may differ in personality as to the extent they want decisions to be surrounded and underpinned by facts and figures. It is possible to define information needs for the successive phases of strategic decision making; this is what many strategy textbooks attempt to do. However, managers have their own styles from which they do not readily depart. The way managers imbibe information can be a decisive factor in the role of information gathering, including management accounting. A number of them wish to be offered information in a very structured manner. Others prefer to be the one who creates order in the turbulent flows of information that surround them. Before taking decisions, some managers wish to consider as many alternatives as possible while others feel they should take action as soon as possible because they believe it is action that provides the most relevant information (ten Bos, 1997). Others are inclined to postpone decisions by repeatedly requiring more information.

The literature distinguishes various strategic (marketing) planning pitfalls (Piercy and Morgan, 1994; de Kare-Silver, 1997; McDonald, 2008). Such pitfalls may be related to both a lack of effective strategic planning (e.g. because of a lack of knowledge and skills, resistance to change, vested interests, little ownership of the plan, confidence in bottom-up creativity) and an overdose of strategic planning (e.g. because previous plans were not taken seriously; planning has become a time-consuming annual ritual; inadequate means).

Sustainability champions in a company should consider these matters in their efforts to advocate sustainability in general or promote a particular sustainability-led project. When operating in a particular organization, the range of possible hurdles can be downsized to a few identifiable issues that relate to the situation. A successful campaign for sustainability must include a focus on the company's strategic planning and capital budgeting practices (or lack of it) to reduce the obstacles as much as possible.

Community-based business models and accounting

Despite the significant role that large (internationally operating) companies have to play in implementing sustainability, in several ways sustainability is increasingly emerging thanks to community-based initiatives (see also Box 3.1). Some are based on private actions by local civilians or grassroots organizations, while in other cases efforts are co-facilitated and co-created by local or regional authorities. Such initiatives often take the form of a cooperative (e.g. to generate and distribute renewable energy) or a social enterprise that combines various types of value creation (e.g. the sales of locally collected used goods combined with employment opportunities for vulnerable people).

That does not detract from the role of established middle-sized and larger companies in engaging in local and regional initiatives, in particular to realize a sustainable economy. Companies need to be able to engage in competition and cooperation simultaneously. These are conflicting demands but, for many businesses, the art of dealing with them is a necessary competence. Although connections with different organizations can evolve without much strategic intent, working on the nature of external relations can be seen as a significant part of a company's strategic management (de Wit and Meyer, 2010).

There are various conceptual notions which help us understand the need for collaboration between different organizations to achieve sustainability. Here, we discuss three of them:

BOX 3.1 BRIDGING THE GAP BETWEEN REGIONAL PLANNING AND THE CORPORATE PERSPECTIVE

For companies to operate more sustainably in respect of the social and ecological systems that surround them, they need to be adequately informed about what impacts they have on these systems. Among others, both simple BSC approaches and more complex, inclusive BSC approaches can contribute to such information.

At the level of regional planning, both policy makers and the managers of precious landscapes – such as local natural parks – use indicators to plan and report on the results of their stewardship. If done well, such measurements are helpful in the enhancement of the sustainable value of natural resources.

In regional planning, there is a tendency to involve local communities. This can be illustrated by the case of urban lake governance that made use of the Social-Ecological System framework (SES) as presented by Bal (2015), building upon the work of Ostrom (2007). See for a further discussion, Annex A.

---

1   The 'embedded organization perspective'
2   Systems of innovation
3   The highest step in sustainability-led innovation.

The 'embedded organization perspective'

This expresses the view that business is about value creation which brings organizations together and lets them focus on a common goal. In various ways, companies can achieve more by working together (de Wit and Meyer, 2010, ch. 7). In many situations, companies want to develop new products together with their buyers, streamline logistics together with their suppliers or expand the industry's potential together with other producers, link technological standards with other industries and improve employment conditions along with the government (de Wit and Meyer, 2010, ch. 7). In the environmental field, there are all kinds of examples of collaboration in industries or subsectors of industries, such as sharing waste collection and waste management systems, joint efforts to make existing production processes cleaner or more energy saving. Regional collaboration can focus on making an entire region clean, green or otherwise attractive and sustainable.

A system of innovation (SI)

This holds the determinants of innovation processes, which include all major economic, social, political, organizational, institutional and possibly other factors that influence the development, diffusion and usage of innovations (Edquist, 2005). This concept is widely used in innovation research. In the developed countries, SIs have evolved in a largely unplanned manner; this implies that they can be subject to government policies only to a limited extent. Perhaps, government policies can be most effective by focusing on national comparative strengths built up over a long period. Emerging economies may show innovative powers as a result of long-term intensive public policies. Essential elements of learning to innovate are:

1 Competence building (through the education system, leading to human capital)
2 Research and Development (R&D) (carried out by research universities and public research organizations and other organizations and individuals),
3 Innovation itself (mainly in companies through organizational learning; it leads to structural capital) (Edquist, 2005).

SIs can be national, sectoral, regional or local. Competition and market-based transactions (products, services or knowledge) are indispensable parts of an SI, but equally essential are pre-market institutions (such as basic schooling), networking and inter-organizational learning. Innovation cannot be left to the market.

The highest step in sustainability-led innovation

Bessant and Tidd (2015, ch. 4) present a model of sustainability-led innovation with three steps: (1) operational optimization (compliance, efficiency reducing harm); (2) organizational transformation (novel products, services or business models, creating shared value); and (3) systems building (novel products, services or business models that are impossible to achieve alone; innovation extends beyond the firm to drive institutional change; it requires the involvement of multiple actors who have not previously worked together). It is 'systems building' in particular that involves the co-evolution of the broader society if life-changing innovations are given a fair chance (new infrastructure, new knowledge, new professionalism, new business rationales).

The non-market (collaboration) aspects of innovation can become visible in different ways (as was largely demonstrated in this chapter), such as:

1 Collaboration in a particular city or region between various parties facilitated by local governments
2 Collaboration between companies in industrial sectors and supplies chains
3 Collaboration between different knowledge institutes, intermediaries, banks and companies
4 Forms of open innovation and co-creation.

In the area of sustainability, it is necessary to make available quantitative indicators that are acceptable to the various groups involved in an innovation process. This requires solid sustainability accounting approaches and, in concrete cases, reliable professional management accountants.

Leadership for sustainability

There is no doubt that both large and small companies have to play leading roles in driving change for and ensuring sustainability. Edmondson *et al.* (2015) distinguish 'multiplier firms', which are firms that develop, disseminate and facilitate extensive sustainability initiatives for their clients or partners. Their authority for implementing changes for sustainability is based on their experts and allies in developing knowledge and disseminating best practices.

However, sustainability leadership is also relevant at other levels. As Ferdig (2007) explains, leaders are often described as those who inspire a shared vision, build consensus, provide direction and foster change in beliefs and actions among their followers. People look to leaders for guidance, direction and answers. Well-managed change seems to be defined as planned, rational and efficient. However, in today's world, accelerated change, complexity and uncertainty have become the norm in our daily lives (Ferdig, 2007). Therefore, instead of giving direction, sustainability leaders should develop and implement actions in collaboration with others, modifying them when needed to adapt to unforeseen changes in the environment over time. Sustainability leadership can be executed outside formal leadership positions, grounded in a personal ethic that reaches beyond self-interest. The experience of change itself and the dissonance it creates fuel new thinking, discoveries and innovations (Ferdig, 2007). The accounting aspects of this reality full of complexity and paradox is still an open question. Balanced Scorecards and business models as we know them appeal to cause-and-effect relationships, business logic and the bottom line. However, these frameworks follow the strategic vision that defines the values to be created. Accounting can tick that box and then can play the humble role of showing the consequences of that vision, in terms of both what it will bring and what it will take. However, as mentioned before, when strategies become volatile, how can accounting prevent coming too late in the day?

References

Abdelkafi, N. and Täuscher, K. (2016) Business models for sustainability from a system dynamics perspective, *Organization & Environment*, 29(1), pp. 74–96.

Bal, M. (2015) Social-Ecological System Framework. Understanding urban lake governance and sustainability in India. Thesis, Erasmus University Rotterdam, The Netherlands.

Bessant, J. and Tidd, J. (2015) *Innovation and Entrepreneurship*. Third edition. Chichester, UK: John Wiley.

Bos, R. ten (1997) *Strategisch denken. Op zoek naar nieuwe helden*, Zaltbommel: Uitgeverij Thema.

Boxall, P. and Purcell, J. (2011) *Strategy and Human Resource Management*. Third edition. Basingstoke, UK: Palgrave Macmillan.

Dyer, L. and Shafer, R. (1999) Creating organizational agility: implications for strategic human resource management. In Wright, P., Dyer, L., Boudreau, J. and Milkovich, G.

(eds) *Research in Personnel and Human Resource Management* (Suppl. 4: Strategic Human Resource Management in the Twenty-First Century), Stanford,CT and London: JAI Press.

Edmondson, A.C., Haas, M.,Macomber, J. and Zuzul, T. (2015) The role of megaprojects and multiplier firms in leading change for sustainability, In Henderson, R., Gulati, R. and Tushman, M. (eds), *Leading Sustainable Change: An organizational perspective*, Oxford: Oxford University Press, pp. 273–97.

Edquist, C. (2005) Systems of Innovation. Perspectives and challenges'. In Fagerberg, J., Mowery, D.C. and Nelson, R.R. *The Oxford Handbook of Innovation*. Oxford: Oxford University Press.

Ferdig, M.A. (2007) Sustainability leadership: Co-creating a sustainable future', *Journal of Change Management*, 7(1), pp. 25–35.

Figge, F., Hahn, T., Schaltegger, S. and Wagner, M. (2002) The Sustainability Balanced Scorecard – linking sustainability management to business strategy', *Business Strategy and the Environment*, 11(5), pp. 269–84.

Freeman, R.E., Harrison, J.S., Wicks, A.C., Parmar, B.L. and De Colle, S. (2010) *Stakeholder Theory. The State of the Art*. Cambridge, UK: Cambridge University Press.

Kaplan, R.S. and Norton, D.P. (2008) *The Execution Premium – linking strategy to operations for competitive advantage*, Boston, MA: Harvard Business Press, p. 159.

Kare-Silver, M. de (1997) *Strategy in Crisis*. London: MacMillan.

Laasch, O. and Conaway, R.N. (2015) Principles of Responsible Management – glocal sustainability, responsibility, and ethics, Stamford, CT: Cengage Learning.

Lynch, R. (2015) *Strategic Management*, seventh edition, Harlow, UK: Pearson.

Martin, R.L. (2014) Strategy should be uncomfortable, *Harvard Business Review*, 92, April, p. 20.

McDonald, M. (2008) *Malcolm McDonald on Marketing Planning. Understanding planning and strategy*. London/Philadelphia: Kogan Page.

Mintzberg, H. (1990) The Design School: reconsidering the basic premises of strategic management, *Strategic Management Journal*, 11(3), 176–95.

Möller, A. and Schaltegger, S. (2005) The Sustainability Balanced Scorecard as a framework for eco-efficiency analysis, *Journal of Industrial Ecology*, 9(4), pp. 73–83.

Ostrom, E. (2007) Sustainable social-ecological systems: an impossibility? Available at www.mcleveland.org/Class_reading/Ostrom_Sustainable_Socio-Economic_Systems.pdf (accessed 29 January 2018).

Pettigrew, A.M. (2009) Corporate responsibility in strategy, in Smith, N.C. and Lenssen, G. (eds), *Mainstreaming Corporate Responsibility*, Chichester, UK: Wiley.

Piercy, N.F. and Morgan, N.A. (1994) Behavioural planning problems in explaining market planning effectiveness, *Journal of Business Research*, 29, pp. 167-178.

Piercy, N.F. (2017) *Market-led strategic change: a guide to transforming the process of going to market*. Fifth Edition. London, New York: Routledge.

Pruijm, R.A.M. (2010), *Grondslagen van Corporate Governance. Leidraad voor behoorlijk ondernemersbestuur, Eerste druk*. Groningen/Houten: Noordhoff.

Roome, N. and Louche, C. (2016) Journeying toward business models for sustainability: A conceptual model found inside the black box of organisational transformation, *Organization and Environment*, 2016, 29(1), pp. 11–35.

Schaltegger, S., Hansen, E.G. and Lüdeke-Freund, F. (2015) Business models for sustainability: Origins, present research, and future avenues', *Organization & Environment*, 29(1), pp. 3–10.

Strikwerda, H. (2009) Wat is de opdracht van het bestuur van de onderneming?, *Holland Management Review*, 128, November/December, pp. 31–9.

SustainAbility (2014) *Model Behavior. 20 Business Model Innovations for Sustainability*. Brooklyn, NY.

Volmann, T.E. (1996) *The Transformation Imperative. Achieving market Dominance through Radical Change*. Boston, MA: Harvard Business School Press.

Wit, B. de and Meyer, R. (2010) *Strategy. Process, Content, Context. An International Perspective*. Fourth Edition. Andover, UK: Cengage Learning EMEA.

Wolters, T. (2013) *Sustainable Value Creation as a Challenge to Managers and Controllers*. Apeldoorn, The Netherlands: Wittenborg University Press.

Annex A. Bridging regional planning and corporate perspectives

For companies to operate more sustainably in respect of the social and ecological systems that surround them, they need to be adequately informed about what impacts they have on these systems. Among others, both simple BSC approaches and more complex, inclusive BSC approaches can contribute to such information.

At the level of regional planning, both policy makers and the managers of precious landscapes – such as local natural parks – use indicators to plan and report on the results of their stewardship. If done well, such measurements are helpful in the enhancement of the sustainable value of natural resources.

In regional planning, there is a tendency to involve local communities. This can be illustrated by a case of urban lake governance that made use of the Social-Ecological System framework (called SES) as presented by Bal (2015) – building upon the work of Ostrom (2009).

Ostrom is a Nobel prize winning economist who contributed to a better understanding of the relationships between ecological and social systems. The strength of the SES framework lies in a continuous broadening of the set of variables involved so that it encompasses almost every characteristic relevant to the understanding of the functioning of social-ecological systems. To deal with the obvious complex nature of the management of natural resources, participative governance approaches have been encouraged, such as a dialogue between regional planners and representatives of relevant economic sectors. These developments can also be translated into BSC approaches that take the complexity and participative approach to planning into account.

Amongst regional planners, a popular way of classifying the value of the natural resources in a region is based on the functions of an ecosystem which provide economic value to a society (Ostrom, 2007). The SES framework includes instrumental values (leading to economic gain) but ignores the intrinsic values of ecosystems (that have to be protected with the aid of sustainability indicators). Resource Provision Services (such as services deriving from water supplies and wood) are monetized while other services (such as health and scenic values) are difficult to value in monetary terms. For the latter, special qualitative and quantitative indicators are used. This development provokes innovations in management accounting approaches that are referred to as community-based management accounting because of the dialogues between the management of firms and their (local) communities as part of the accounting process. This development is also conducive to meaningful sustainability strategies at different geographical levels. See Figure 3.1

Such a framework can be helpful in developing a common understanding of the values involved in certain economic activities and the use they make of the community's resources. That understanding can serve the parties involved to engage in common investments in the region's resources and sustainable ways of producing goods and services.

Existing methods of resource accounting and monetizing external effects can play a role in clarifying what the 'true' costs and benefits of economic activities – beyond conventional costs and benefits – are and how these are divided among the various community segments. In the context of individual enterprises, governmental agencies often account for the natural resources they manage. For some natural resources, the participation of organizations with an interest in natural resources are mediated by markets. Others are managed by governments and offered as public goods. In this context, both regional planners and companies

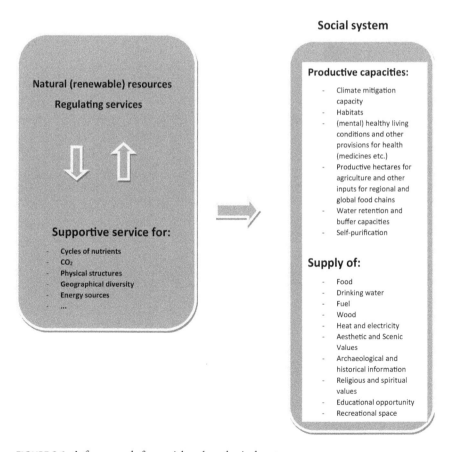

FIGURE 3.1 A framework for social and ecological systems

can make use of regional accounting data about the performance of social and ecological systems.

To make the best use of these accounting data, a BSC approach could be used which combines the business-economic perspective (involving indicators in the areas of finance, customer relations, internal organizational and processes and the supporting learning and growth capacities) with a broader natural resource perspective. The indicators that measure this broader natural resource perspectives (and, in particular, the sustainability aspects attached to them) should be based on a dialogue with the company's stakeholders, especially with those who are directly affected by or concerned about the company's impacts on the various ecosystems. Indicator techniques that could be used within this context are the measurement of ecological footprints and external labelling and certification procedures with regard to the inputs, throughput and output of companies. Although, to a certain extent, the interests of the various parties will be opposed to each other, by agreeing on a common set of indicators, regional planners, companies and civil organizations can reach a kind of cooperation that is beneficial to all stakeholders in a particular region and to the society at large.

4

SUSTAINABILITY IN PROJECT MANAGEMENT

Making it happen

A.J. Gilbert Silvius

Introduction

When the late sustainability icon Ray Anderson was asked about the role of project management in sustainability, his answer was 'It's the role of implementing new strategies, it's the role of making it happen, it's the role of turning ideas and aspirations into reality' (Anderson, 2009). With those words he highlighted the relationship between project management and sustainability. Projects are 'instruments of change' (Silvius *et al.*, 2012) that implement new strategies and ambitions in organizations. When organizations adopt a more sustainable strategy, projects inevitably play a key role in the transition to a sustainable enterprise. Quite literally, projects and their management are 'the way to sustainability' (Marcelino-Sádaba *et al.*, 2015).

This relationship between sustainability and project management, however, is surrounded by challenges. Projects are defined by their temporary nature (Lundin and Söderholm, 1995; Turner and Müller, 2003) and the resulting short-term, task-oriented perspective of project management does not seem to be compatible with the long-term perspective of sustainability (Gareis *et al.*, 2013). Projects and sustainable development are therefore not "*natural friends*" (Silvius *et al.*, 2012). Nevertheless, it can be observed that the integration of sustainability and project management is 'picking up momentum' (Silvius and Tharp, 2013, xix) while sustainability is being recognized as a developing theme in project management research (Silvius, 2017). In fact, several literature reviews on the topic (for example, Aarseth *et al.*, 2017; Marcelino-Sádaba *et al.*, 2015; Silvius and Schipper, 2014) report a significant growth in relevant publications over the last 10 years.

With the relationship between sustainability and projects emerging, the question arises how considering sustainability influences project management. It is this question that this chapter aims to answer.

Background

Projects as instruments of change

The Project Management Institute, the most influential professional organization of project managers, defines a project as 'A temporary endeavor undertaken to create a unique product, service, or result' (Project Management Institute, 2013: 553). This definition emphasizes the temporary aspects of a project and its goal-oriented nature. However, there is more to projects than just the intended result. The ISO 21500 standard on project management, for example, defines a project as 'a unique set of processes consisting of coordinated and controlled activities, with start and end dates, performed to achieve project objectives' (ISO 21500). This definition also mentions the goal orientation but defines a project as a set of processes and activities. A third industry standard, PRojects In Controlled Environments, version 2 (PRINCE2) adds an organizational component, by defining a project as 'A temporary organization that is created for the purpose of delivering one or more business products according to an agreed Business Case' (PRINCE2, 2009:3). Reflecting on these definitions, it can be concluded that a project is a temporary venture, existing of a set of non-routine processes or activities, performed by a temporary organization, in order to deliver a product, service or result that serves a purpose.

The purpose of a project is a frequently debated topic; different words are used in this context (purpose, goal, objective, deliverable, output, outcome, etc.) and it is not always clear what is indicated by these. Silvius *et al.* (2012) have developed the following model to position projects. In their view, projects are, as temporary organizations, connected to a non-temporary 'permanent' organization. Figure 4.1 illustrates this relationship.

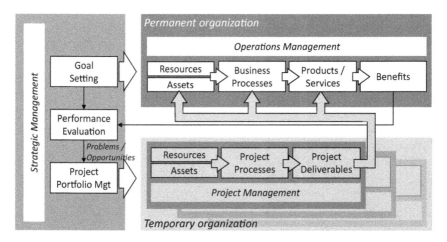

FIGURE 4.1 Projects as temporary organizations that deliver change to permanent organizations

The permanent organization utilizes resources and assets in its business processes to deliver products or services to its customers in order to generate benefits or value to the organization and its stakeholders. Its activities are based on goals that are developed or set in a strategic management process.

The benefits that the organization realizes are evaluated against the set goals. If the performance is satisfactory, the business operations can continue. However, if the performance is unsatisfactory, or new opportunities arise, there may be reason to carry through some changes. In that case, a temporary organization, in the form of a project, is commonly used to create this change. The change may concern the resources and assets of the permanent organization, but also the products/services rendered or the internal policies and procedures. This change perspective on projects is also found in Turner (2014), who defines project management as '*the means by which the work of the resources assigned to the temporary organization is planned, managed and controlled to deliver the beneficial change*' (Turner, 2014, 29).

The change perspective on projects provides new input on the purpose of projects. Projects are not justified by the deliverable or product they realize, but by the change in the permanent organization this deliverable invokes and the benefits this change realizes. And although different standards and languages are not always clear on this difference, the words 'output' and 'outcome' seem to capture this distinction quite well. The *output* of the project is the measurable artefact the project delivers at the end of its lifecycle, whereas the *outcome* of the project is the effect and benefits of the changes this artefact causes in the permanent organization.

Projects and sustainability

The change aspects of sustainability was already highlighted in the seminal Brundtland report, that stated "In essence, sustainable development is a process of change in which the exploitation of resources, the direction of investments, the orientation of technological development and institutional change are all in harmony and enhance both current and future potential to meet human needs and aspirations" (World Commission on Development and Environment, 1987). Combining the change perspective on projects and the requirement of change that sustainability entails, Marcelino-Sádaba *et al.* (2015) observe that "projects are the ideal instrument for change" and "the necessary change that we require towards sustainability will be boosted by applying the project management discipline to sustainab*ility*" (Marcelino-Sádaba *et al.*, 2015).

As mentioned in the Introduction, the concepts of sustainability and sustainable development are increasingly also being related to project management (for example, by Labuschagne and Brent, 2005, 2007; Edum-Fotwe and Price, 2009; Maltzman and Shirley, 2011; Silvius *et al.*, 2012; Gareis *et al.*, 2013; Martens and Carvalho, 2016). Silvius (2017) identifies four defining characteristics of the 'sustainability' school of project management: considering projects in a societal perspective, adopting a management for stakeholders approach, considering all aspects of the projects through the triple bottom-line perspectives of economic, environmental and social interests and adopting a value-based approach to project management. The following section discusses these characteristics of sustainable project management further.

Projects in a societal perspective

As suggested by Marcelino-Sádaba *et al.* (2015), the sustainability school adopts a societal perspective on projects and considers projects as instruments to realize societal change. In their literature review on publications on sustainability in project management, Silvius and Schipper (2014: 72) therefore identify the 'recognition of the context of the project' as the starting point of considering sustainability in project management. "Integrating the dimensions of sustainability in project management inevitably implies a broader consideration of the context of the project" (Silvius and Schipper, 2014, 72). A recent study into the 'projectification' of three Western European countries showed that projects account for roughly one third of economic activity (Schoper *et al.*, 2018), which justifies this societal perspective. However, the role of projects in society is not limited to economic value. The Sustainability school elaborates on this societal role by considering also the social and environmental impact of projects.

Silvius *et al.* (2017) illustrate this broadened perspective on the change that projects realize with Figure 4.2.

The broadened contextual orientation illustrated in Figure 4.2 builds upon the interacting life cycles as proposed by Labuschagne and Brent (2005), and combines this concept with a wider stakeholder orientation which includes a reference to the (local and global) society.

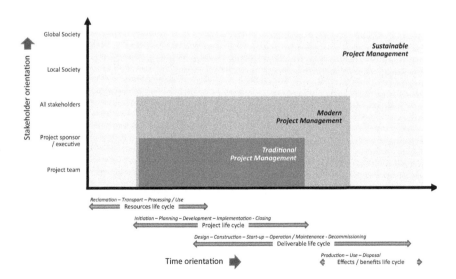

FIGURE 4.2 The broadened scope of sustainable project management

Silvius *et al.*, 2017

Management for stakeholders

Several authors (AlWaer *et al.*, 2008; Eskerod and Huemann, 2013; Labelle and Leyrie, 2013) recognize the need for a more proactive engagement of stakeholders as a consequence of integrating sustainability into project management. Eskerod and Huemann (2013) conclude that the current standards of project management guide practitioners towards the recognition of a rather limited group of stakeholders and to "selling the project to the most important stakeholders rather than involving them and their interests into the creation of project objectives" (Eskerod and Huemann, 2013, 43). In this approach, the 'management-of-stakeholders' approach, stakeholders are seen primarily as providers of resources that should be prevented from hindering the project. "In contrast, the 'management-for-stakeholders' approach (Freeman *et al.*, 2007, 2010) takes the point of departure that all stakeholders have the right and legitimacy to receive management attention (Julian *et al.*, 2008). Stakeholders are not means to specific aims in the organization but valuable in their own rights" (Eskerod and Huemann, 2013, 40). According to Labelle and Leyrie (2013), this implies that stakeholder communication becomes stakeholder participation. The information flow between project and stakeholders is no longer unidirectional but transformed into a dialogue that allows participants to take part in developing the project (Libaert, 1998).

Triple bottom-line criteria for all aspects of the project

Integrating sustainability in project management will influence the specifications and requirements of the project's deliverable or output, and the criteria for project success (Eid, 2009; Maltzman and Shirley, 2011) – for example, the inclusion of environmental or social aspects in the project's objective and intended output and outcome (Silvius *et al.*, 2012). Integrating sustainability into project management suggests that the content, intended output/outcome and success criteria are based on a holistic view of the project (Gareis *et al.*, 2013) and developed together with a broad group of stakeholders (Eskerod and Huemann, 2013). Integrating sustainability also implies that the definition and perception of project success take into account the 'triple bottom line' of economic, social and environmental benefits as laid out in the business case, both in the short term as in the long term (Silvius and Schipper, 2014). This implies therefore, that the success of the project is to be assessed based on the life cycle of the project and its outcome (Pade *et al.*, 2008; Craddock, 2013).

Values based

Sustainability is inevitably a normative concept, reflecting values and ethical considerations of society. "Sustainability is the ideal state of sustainable development efforts" (Keeys and Huemann, 2017), which is based on the ethics and values of the actors. Following the conclusion that sustainability is embedded in the values of the social system to which the sustainability relates, a logical question is which values sustainability is based upon. In the Brundtland commission's definition of sustainable development, the statement ". . . the needs of the present without compromising the ability of future generations . . ." (World Commission on Environment and Development, 1987) implies equality as a value of sustainability. In the definition, equality is applied to the rights of different generations, but the value may also be applied to the interests of different stakeholders. This interpretation can be found with the previously mentioned stakeholder theory. Other values associated with sustainability are participation, fairness, respect, honesty, transparency and traceability.

Dimensions of sustainability.

Although the sustainability school of thinking may be united in the characteristics described above, different authors and studies still differ in focus on the implications of sustainability in projects and project management. Based on a study of 164 publications on the topic, Silvius and Schipper (2014) concluded that the following dimensions of sustainability are relevant to the understanding of sustainability in the context of project management.

Sustainability is about balancing or harmonizing social, environmental and economic interests

In order to contribute to sustainable development, a company should satisfy all 'three pillars' of sustainability: social, environment and economic (Elkington, 1997). The dimensions are interrelated – that is, they influence each other in various ways.

Sustainability is about both short- and long-term orientation

A sustainable company should consider both short-and long-term consequences of their actions, and focus not only on short-term gains (Gareis *et al.*, 2011). The dimension of both short- and long-term orientation focuses the attention on the full lifespan of the matter in hand (Brent and Labuschagne, 2006).

Sustainability is about local and global orientation

The increasing globalization of economies affects the geographical area influenced by organizations. Intentionally or not, many organizations are influenced by international stakeholders whether these are competitors, suppliers or (potential) customers. The behaviour and actions of organizations therefore have an effect on economic, social and environmental aspects, both locally and globally. 'In order to efficiently address these nested and interlinked processes, sustainable development has to be a coordinated effort playing out across several levels, ranging from the global to the regional and the local' (Gareis *et al.*, 2011, p. 61).

Sustainability is about values and ethics

As argued by Robinson (2004) and Martens (2006), sustainable development is inevitably a normative concept, reflecting values and ethical considerations of society. Part of the change needed for more a sustainable development will, therefore, also be the implicit or explicit set of values that we as professionals, business leaders or consumers have and that influence or lead our behaviour.

Sustainability is about transparency and accountability

The principle of transparency implies that an organization is open about its policies, decisions and actions, including the environmental and social effects of those actions and policies (International Organization for Standardization, 2010). This implies that organizations provide timely, clear and relevant information to their stakeholders so that they can evaluate the organization's actions and can address potential issues with these actions.

Complementing the principle of transparency is the principle of accountability. This principle implies that an organization is responsible for its policies, decisions and actions and their effects on environment and society. The principle also implies that an organization accepts this responsibility and is willing to be held accountable for these policies, decisions and actions.

Sustainability is about stakeholder participation

Considering and respecting the potential interests of stakeholders is key to sustainability. ISO 26000 emphasizes the behavioral side of this principle, by mentioning 'proactive stakeholder engagement' as one of its principles (International Organization for Standardization, 2010). Stakeholder participation therefore requires "a process of dialogue and ultimately consensus-building of all stakeholders as partners who together define the problems, design possible solutions, collaborate to implement them, and monitor and evaluate the outcome" (Goedknegt and Silvius, 2012, 3).

Sustainability is about risk reduction

The so-called precautionary principle is based on the understanding that in environment–society system interactions, the complexity, indeterminacy, irreversibility and nonlinearity has reached a level at which it is more efficient to prevent damage rather than ameliorate it (Turner, 2010). The 2010 Deepwater Horizon oil-spill disaster has fuelled the discussion on the suitability of financial risk management techniques for societal and environmental risks.

Sustainability is about eliminating waste

The importance of eliminating waste is mentioned by several authors, including Maltzman and Shirley (2011, 2013). They refer to 'The Seven Wastes' as identified in the Toyota production system. These seven wastes are: overproduction, waiting, transporting, inappropriate processing, unnecessary inventory, unnecessary or excess motion, and defects.

The principle of eliminating waste can also be found in the cradle-to-cradle concept (McDonough and Braungart, 2002) that builds upon the idea that waste equals food.

Sustainability is about consuming income, not capital

Sustainability implies that nature's ability to produce or generate resources or energy remains intact. The 'source and sink' functions of the environment should not be degraded, meaning that the extraction of renewable resources should not exceed the rate at which they are renewed, and the absorptive capacity of the environment to assimilate waste should not be exceeded (Gilbert *et al.*, 1996). This principle may also be applied to the social perspectives (Silvius *et al.*, 2012). Organizations should also not 'deplete' people's ability to produce or generate labour or knowledge by physical or mental exhaustion. In order to be sustainable, companies have to manage not only their economic capital, but also their social and environmental capital.

Sustainable project management

The nine dimensions of sustainability discussed above provide guidance to the integration of the concepts of sustainability into project management. This consideration of sustainability has led to the emerging concept of 'Sustainable Project Management', defined as "the planning, monitoring and controlling of project delivery and support processes, with consideration of the environmental, economic and social aspects of the life-cycle of the project's resources, processes, deliverables and effects, aimed at realizing benefits for stakeholders, and performed in a transparent, fair and ethical way that includes proactive stakeholder participation" (Silvius and Schipper, 2014, 79).

From the emerging literature on the integration of sustainability and project management, two types of relationship between sustainability and project management appeared (Silvius and Schipper, 2015; Kivilä *et al.*, 2017): the sustainability of the project's *product* (the deliverable that the project realizes) and the sustainability of the project's *process* of delivering and managing the project.

Content-related approaches

The dimensions of sustainability discussed in the previous section provide input for integrating sustainability requirements into the content-related aspects of the project, such as the specifications and design of the project's deliverable (Eid, 2009; Aarseth *et al.*, 2017), materials used (Akadiri *et al.*, 2013), benefits to be achieved (Silvius *et al.*, 2012), business case (Weninger and Huemann, 2013) and quality and success criteria (Martens and Carvalho, 2016). Studies on the integration of sustainability into project management that take this content related perspective, often focus on operationalizing the Triple Bottom Line concept by developing sets of indicators on the different perspectives (For example Bell and Morse, 2003; Keeble *et al.*, 2003; Labuschagne and Brent, 2006; Fernández-Sánchez and Rodríguez-López, 2010; Martens and Carvalho, 2016). Considering sustainability in these aspects will mostly result is a more sustainable project in terms of a more sustainable deliverable; however, this approach carries the risk of lacking the holistic approach of the integration of the economic, environmental and social perspectives.

Process-related approaches

Some studies focus on the integration of the dimensions of sustainability into the processes of project management and delivery, such as the identification and engagement of stakeholders (Eskerod and Huemann, 2013), the process of procurement in the project (Molenaar and Sobin, 2010), the identification and management of project risks (Silvius, 2016), the communication in and by the project (Pade *et al.*, 2008) and the selection and organization of the project team (Silvius and Schipper, 2014). Gareis *et al.* (2013) observe that this perspective has received less attention than the content-oriented perspective. A potential explanation for this is the temporary nature of projects (Gareis *et al.*, 2013). This temporariness of projects may lead to the view that the sustainability, or unsustainability, of the project's process is less impactful. However, Labuschagne and Brent (2005), point out that in the process of developing and delivering a project, also many content-related aspects are decided and that therefore a project's process and product interact.

In the study by Eid (2009), a forum of project management practitioners was asked about their assessment of the impact of sustainable development on project management processes. More specifically, for each project management process group (initiating, planning, executing, controlling, closing) the study asked their views on the area of integration of sustainability aspects. The questions asked were:

- To which account do the scope and objectives of the project (more or less the project content) provide opportunities for integrating sustainability?
- To which account do the actual processes of delivering and managing the project (the project process) provide opportunities for integrating sustainability?

Figure 4.3 shows the result of Eid's study. This study shows there are opportunities for the integration of sustainability in all process groups of project management. The area of integration of sustainability aspects, however, differs. The initiating and planning processes of the project provide opportunities for

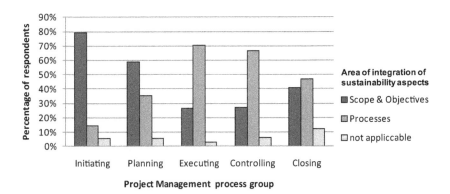

FIGURE 4.3 The best areas to integrate sustainable development into project management

Based on Eid (2009)

integrating sustainability into the content of the project. The executing and controlling processes of the project provide opportunity for integrating sustainability in the process of the project.

A more specific assessment of the possible impact of sustainability on project management is presented in the following section, which discusses the consideration of sustainability in a number of 'impact areas'.

Recognition of the context of the project

As illustrated in Figure 4.2, a starting point for the integration of sustainability in project management is the recognition of the context of the project. Integrating the dimensions of sustainability in project management inevitably implies a broadening of a project's context that is taken into consideration (Silvius *et al.*, 2012; Tharp, 2013). Both the time and the spatial boundaries of the context are stretched when considering sustainability. In particular, the sustainability dimensions 'both short and long term' and 'local and global' impact the project context. In an increasingly global business world, more and more projects also touch upon the geo-economic challenges as part of the project team may be located in emerging economies like India or China, and suppliers may be anywhere in the world or customers benefitting from the project's deliverable. It is clear that the globalizing business world also includes globalized projects and project management. Within the project management community, the discussion about globalization aspects of the result or deliverable of the project still has to emerge.

Identification of stakeholders

The dimensions of sustainability, more specifically the principles 'balancing or harmonizing social, environmental and economical interests', 'both short term and long term' and 'both local and global', will likely increase the number of stakeholders of the project (Eskerod and Huemann, 2013; Tharp, 2013). Typical 'sustainability stakeholders' may be environmental protection pressure groups, human rights groups, nongovernmental organizations, etc. (Silvius *et al.*, 2012).

Project specifications/requirements/deliverable/quality criteria

Integrating the principles of sustainability will influence the specifications and requirements of the project's deliverable or output, and the criteria for the quality of the project (Eid, 2009; Maltzman and Shirley, 2010; Taylor, 2010) – for example, the inclusion of environmental or social aspects in the project's objective and intended output and outcome (Silvius *et al.*, 2012).

Integrating sustainability into project management suggests that the content, intended output/outcome and quality criteria are based on a holistic view of the project (Gareis *et al.*, 2013), including sustainability dimensions as 'economic environmental and social', 'short term and long term' and 'local and global', and developed together with a broad group of stakeholders (Eskerod and Huemann, 2013).

Business case/costs/benefits

The influence of the principles of sustainability on the project content will need to be reflected also in the project justification (Silvius, 2015). The identification of costs, benefits and the business case of the project may need to be expanded to include also non-financial factors that refer, for example, to social or environmental aspects (Gareis *et al.*, 2011, 2013).

Considering the concepts of sustainability in project management implies that the business case of a project addresses the triple bottom line of economic, social and environmental benefits. Silvius *et al.* (2012) conclude that this implies a multi-criteria approach to investment evaluation, with consideration of both financial and non-financial criteria.

Dimensions of project success

Shenhar *et al.* (2001) relate the criteria and perception of project success to a time scale. In their model, the perception of success develops from the short-term criterion of 'project efficiency' to the medium-term criterion of 'business success', and eventually the long-term criterion of 'prepare for the future'. Given the sustainability dimension 'short term and long term', it has to be concluded that integrating sustainability into the dimensions implies also that criteria which refer to the development of future capabilities of the organization are included in the dimensions of project success, next to the traditional 'iron triangle' or project efficiency. This also implies that the success of the project is assessed based on the life cycle of the project and its outcome (Pade *et al.*, 2008; Craddock, 2013).

Taking into account the sustainability dimension 'economic, environmental and social' implies that the criteria of project's success refer to economic, social and environmental benefits.

A conceptual mapping of the different relationships between dimensions of sustainability and criteria of project success revealed that, for most relationships between sustainability and success criteria, it is expected that the consideration of sustainability increases the likelihood of success (Silvius and Schipper, 2016).

Selection and organization of the project team

Another area of impact of sustainability is the project organization and management of the project team. In particular, the social aspects of sustainability, such as equal opportunity and personal development, can be put to practice in the management of the project team (Tharp, 2013), but also aspects such as commuting distance and work–life balance may be considered in the organization and management of the team.

Van den Brink (2013) applies the principles of positive psychology to project management in order to manage the project team in a sustainable way.

Project sequencing and schedule

Taylor (2010) recognizes the opportunities for considering sustainability in project planning, scheduling and sequencing. He challenges project managers to think beyond 'how things are normally done', and provides several examples, one of the examples being off-site fabrication, rather than on-site. This provides potential sustainability advantages of less waste, reduced delivery costs, better use of resources, opportunities to increase labour skills, opportunities for job creation in poorer locations, economies of mass production, etc.

Sustainable project management also implies performing the project as efficient as possible, thereby minimizing waste. Waste can occur in materials, but also in idle resources or waiting times (Maltzman and Shirley, 2011). These last two types of waste typically relate to scheduling and sequencing.

Materials used

An obvious impact area for sustainability in project management is the selection of materials used in the project (Silvius *et al.*, 2012; Akadiri *et al.*, 2013). Logical considerations are the use of hazardous substances, pollution and energy use, both in the production process and in the use and remaining life cycle of the materials. However, the most important sustainability insight regarding materials might be applying a life cycle perspective (Brent and Petrick, 2007). This implies considering the production supply chain, but also durability, reusability and recyclability at the decommission stage of the project's deliverable.

Procurement

Not just the materials used, but also the processes concerned with procurement and selection of suppliers, provide a logical opportunity to integrate considerations of sustainability. For example, appreciating the sustainability performance of potential suppliers in supplier selection (Taylor, 2010) but also in fighting bribery and non-ethical behaviour in procurement (Tharp, 2013), both by participants in the project or organization and by potential suppliers or authorities.

Risk identification and management

Risk management, including risk mitigation, is a well-known concept in project management. According to project management standards, a risk is defined as an uncertain event or set of events that, should it occur, will have an effect on the achievement of objectives (Office of Government Commerce, 2010). However, when considering this definition from a sustainability perspective, some questions may arise. For example, which objectives will be considered in the identification of potential risks? Are these the objectives of the projects or the desired outcomes of the project's deliverable? And whose objectives will be considered? The objectives of the project sponsor or the objectives of all stakeholders?

With the inclusion of the concepts of sustainability in project management, the assessment of potential risks will need to evolve (Winnall, 2013). Logically in the identification of risks, also environmental and social risks are to be considered. And, following the life cycle approach, these risks need to be assessed for the project's resources, processes, deliverables and effects (Silvius *et al.*, 2012). However, considering sustainability in risk identification and management does not only apply to the kind of risks considered. It also implies that risks are considered from the different points of views and interests of all stakeholders, not just the project sponsor. This also suggests that in sustainable project management, the stakeholders are participating in the identification, assessment and management of risks (Silvius and Tharp, 2013).

Stakeholder participation

Several authors (e.g. Perrini and Tencati, 2006; Pade *et al.*, 2008; Gareis *et al.*, 2009) emphasize the importance of stakeholder participation in projects. This principle logically impacts the stakeholder management and the communication processes in project management. However, the intention behind 'participation' goes beyond the identification of some specific processes. Stakeholder participation is not so much a specific process, as it is an attitude with which all project management processes are performed.

According to the ISO 26000 guideline, proactive stakeholder engagement is one of the basic principles of sustainability (International Organization for Standardization, 2010). Also Eskerod and Huemann (2013) link sustainable development, projects and the role of stakeholders, and conclude that there is a need 'to incorporating stakeholders and their interests in more project management activities' (Eskerod and Huemann, 2013: 45).

Project communication

Following the principle of transparency and accountability, incorporating sustainability into project management processes and practices would imply proactive and open communication about the project, that would also cover social and environmental effects, both short- and long-term (Khalfan, 2006; Taylor, 2010; Silvius *et al.*, 2012).

Project reporting

As the project progress reports logically follow the definition of scope, objective, critical success factors, business case, etc., so too will the project reporting processes be influenced by the inclusion of sustainability aspects (Perrini and Tencati, 2006).

Project handover

Various studies show a more diverse picture of the opportunities available to integrate sustainability aspects. For example, the respondents in Eid's study (2009) suggest that the closing phase of a project offers the least appealing opportunities for integrating sustainability (Eid, 2009). However, Pade *et al.* (2008) and Silvius *et al.* (2012) point out the importance of also the closing processes for a more sustainable project result. The closing processes typically include handover to the permanent organization. The success of this handover and the acceptance of the project result are important aspects of a project's sustainability. Failed or non-accepted projects can hardly be considered sustainable, given the waste of resources, materials and energy they represent.

Organizational learning

A final area of impact of sustainability is the degree to which the organization learns from the project. Sustainability also suggests minimizing waste. Organizations should therefore learn from their projects in order to not 'waste' energy, resources and materials on their mistakes in projects (Eid, 2009; Silvius *et al.*, 2012).

Reflection

The preceding section identified the impact of the dimensions of sustainability on project management. Considering these impacts provides new or additional perspectives on the content and the process of the project. Several authors conclude that integrating sustainability requires a scope shift in the management of projects: from managing time, budget and quality to managing social, environmental and economic impacts (Haugan, 2012; Silvius *et al.*, 2012; Ebbesen and Hope, 2013).

However, the impact of sustainability on project management is more than adding a new perspective or aspect to processes and formats of the current project management standards. Adding new perspectives to the way projects are considered also adds complexity (Silvius *et al.*, 2012; Eskerod and Huemann, 2013). Project management therefore needs a more holistic and less mechanical approach (Gareis *et al.*, 2013). The traditional project management paradigm of controlling time, budget and quality suggests a level of predictability and control that does not apply to changes that are considered in a global and long-term perspective. Most likely, these changes and their effects are not completely known and overseeable. The integration of sustainability therefore requires a paradigm shift (Silvius *et al.*, 2012) – from an approach to project management that can be characterized by predictability and controllability of both process and deliverable, to an approach that is characterized by flexibility, complexity and opportunity (Silvius and Schipper, 2014).

The basis for the scope shift and the paradigm shift described above is the way the project management professional sees him/herself (Crawford, 2007). Traditionally, project managers tend to serve their project sponsors and 'do what they are told' to do. They position themselves as subordinate to the project sponsor and manage their project team around scope, stakeholders, deliverables, budget, risks and resources as specified by the stakeholder's requirements. However, project managers are well positioned to play a significant role in the implementation of the concepts of sustainability in organizations and business (Association for Project Management, 2006; Tharp, 2012). Taking up this responsibility changes the role of project managers and therefore changes the profession (Silvius *et al.*, 2012). Integrating sustainability requires that project managers develop themselves as specialists in sustainable development and act as partner of and peer to stakeholders (Crawford, 2007; Tam, 2010). In this mind-shift, the change a project brings about is no longer a given for which only the project sponsor can be held responsible: it is also the responsibility of the project manager with ethics and transparency as essential touchstones. Project management is no longer about 'managing' stakeholders, but about engaging with stakeholders in realizing a sustainable organization and society.

The three shifts identified above can be summarized and connected as illustrated in Figure 4.4. The scope shift resides in the paradigm shift of sustainable project management. This paradigm shift is grounded in the mind-shift (Silvius and Schipper, 2014).

Conclusion

Projects can make a contribution to an organization's sustainability. It can, therefore, be expected that projects and the project management standards reflect the concepts and principles of sustainability. Based on the project management literature, this chapter identified the areas of impact of sustainability on project management. These areas of impact are: recognition of the context of the project; identification of stakeholders; project specifications/requirements/deliverable/quality criteria;

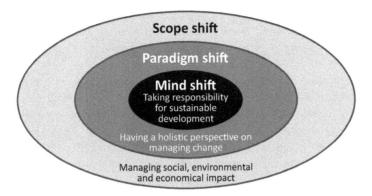

FIGURE 4.4 The three shifts of sustainable project management

Silvius and Schipper (2014)

business case/costs/benefits; dimensions of project success; selection and organization of project team; project sequencing and schedule; materials used; procurement; risk identification and management; stakeholder involvement; project communication; project reporting; project handover; and organizational learning. In short, we conclude that considering sustainability has an impact on practically all processes and practices of project management.

Although this list of areas of impact provides a comprehensive overview of the aspects of project management that are affected by the consideration of sustainability, it does not cover the full depth of the concepts of sustainability, and the pivotal role of the project manager in realizing sustainable development in organizations. In addition to the areas of impact, we concluded three shifts that characterize the integration of sustainability and project management. Considering sustainability implies firstly a shift of scope in the management of projects: from managing time, budget and quality to managing social, environmental and economic impact. Secondly, it implies a shift of paradigm of project management: from an approach that can be characterized by predictability and controllability, to an approach that is characterized by flexibility, complexity and opportunity. And thirdly, considering sustainability implies a mind-shift for the project manager: from delivering requested results, to taking responsibility for sustainable development in organizations and society.

With these findings, the practices and standards of project management can be developed further to address the role played by projects in creating a sustainable organization.

References

Aarseth, W., Ahola, T., Aaltonen, K., Økland, A. and Andersen, B. (2017) Project sustainability strategies: A systematic literature review, *International Journal of Project Management*, 35(6), pp. 1071–83.

Akadiri, P.O., Olomolaiye, P.O. and Chinyio, E.A. (2013) Multi-criteria evaluation model for the selection of sustainable materials for building projects, *Automation in Construction*, 30, pp. 113–25.

AlWaer, H., Sibley, M. and Lewis, J. (2008) Different stakeholder perceptions of sustainability assessment', *Architectural Science Review*, 51(1), pp. 48–59.

Anderson, R. (2009), Ray Anderson on project managers and sustainability', retrieved from www.youtube.com/watch?v=bnUbuginMGE

Association for Project Management (2006) *APM supports sustainability outlooks*.

Bell, S. and Morse, S. (2003) *Measuring Sustainability Learning from Doing*. London: Earthscan.

Brent, A.C. and Labuschagne, C. (2006) Social indicators for sustainable project and technology life cycle management in the process industry, *International Journal of Life Cycle Assessment*, 11(1), pp. 3–15.

Brent, A.C. and Petrick, W. (2007) Environmental Impact Assessment (EIA) during project execution phases: Towards a stage-gate project management model for the raw materials processing industry of the energy sector, *Impact Assessment and Project Appraisal*, 25(2), pp. 111–22.

Craddock, W.T. (2013) How business excellence models contribute to project sustainability and project success, in Silvius A.J.G. and Tharp, J. (eds), *Sustainability Integration for Effective Project Management*, Hershey, PA: IGI Global Publishing.

Crawford, L. (2007) Global body of project management knowledge and standards', in Morris, P.W.G. and Pinto, J.K. (eds), *The Wiley Guide to Managing Projects*, Hoboken, NJ: Wiley.

Ebbesen, J.B. and Hope, A.J. (2013) Re-imagining the Iron Triangle: Embedding sustainability into project constraints', *PM World Journal*, 2(3), pp. 1–13.

Edum-Fotwe, F.T. and Price, A.D.F. (2009) A social ontology for appraising sustainability of construction projects and developments, *International Journal of Project Management*, 27(4), pp. 313–22.

Eid, M. (2009) *Sustainable Development & Project Management*. Cologne, Germany: Lambert Academic Publishing.

Elkington, J. (1997) *Cannibals with Forks: the Triple Bottom Line of 21st Century Business*, Oxford: Capstone Publishing.

Eskerod, P. and Huemann, M. (2013) Sustainable development and project stakeholder management: what standards say, *International Journal of Managing Projects in Business*, 6(1), pp. 36–50.

Fernández-Sánchez, G. and Rodríguez-López, F. (2010) A methodology to identify sustainability indicators in construction project management—Application to infra-structure projects in Spain, *Ecological Indicators*, 10, pp. 1193–201.

Freeman, R.E., Harrison, J.S. and Wicks, A.C. (2007) *Managing for Stakeholders: Survival, reputation, and success*, Yale, CT: Yale University Press.

Gareis, R., Huemann, M. and Martinuzzi, A. (2009) *Relating Sustainable Development and Project Management*, Berlin: IRNOP IX.

Gareis, R., Huemann, M. and Martinuzzi, A. (2011) What can project management learn from considering sustainability principles?, in *Project Perspectives*, XXXIII, pp. 60–5, International Project Management Association.

Gareis, R., Huemann, M. and Martinuzzi, R.-A., with the assistance of Weninger, C. and Sedlacko, M. (2013) *Project Management & Sustainable Development Principles*, Newton Square, PA: Project Management Institute.

Gilbert, R., Stevenson, D., Girardet, H. and Stern, R. (eds) (1996) *Making Cities Work: The role of local authorities in the urban environment*, London: Earthscan.

Goedknegt, D. and Silvius, A.J.G. (2012), The implementation of sustainability principles in project management, Proceedings of the 26th IPMA World Congress, Crete, pp. 875–82.

Haugan, G. (2012) *The New Triple Constraints for Sustainable Projects, Programs, and Portfolios*, Boca Raton, Fl: CRC Press.

International Organization for Standardization (2010) *ISO 26000, Guidance on Social Responsibility*, Geneva.

International Organization for Standardization (2012) *ISO 21500:2012, Guidance on Project Management*, Geneva.

Keeble, J.J., Topiol, S. and Berkeley, S. (2003), Using indicators to measure sustainability performance at a corporate and project level, *Journal of Business Ethics*, 44(2–3), 149–58.

Keys, L.A. and Huemann, M. (2017), Project benefits co-creation: Shaping sustainable development benefits, *International Journal of Project Management*, 35(6), 1196–212.

Khalfan, M.M.A. (2006), Managing sustainability within construction projects, *Journal of Environmental Assessment Policy and Management*, 8(1), pp. 41–60.

Kivilä, J., Martinsuo, M. and Vuorinen, L. (2017), Sustainable project management through project control in infrastructure projects, *International Journal of Project Management*, 35(6), 1167–83.

Labelle, F. and Leyrie, C. (2013), Stakepartner management in projects, *The Journal of Modern Project Management*, May–August.

Labuschagne, C. and Brent, A. C. (2005), Sustainable project life cycle management: The need to integrate life cycles in the manufacturing sector, *International Journal of Project Management*, 23(2), pp. 159–68.

Labuschagne, C. and Brent, A. C. (2007), Sustainability assessment criteria for projects and technologies: Judgements of industry managers, *South African Journal of Industrial Engineering*, 18(1), pp. 19–33.

Libaert, T. (1998). Faire accepter un projet: principes et méthodes, *Communication et langages*, 117, pp. 76–90.

Lundin R.A. and Söderholm A. (1995) A theory of the temporary organization, in *Scandinavian Journal of Management*, 11, pp. 437–55.

Maltzman, R. and Shirley, D. (2011) *Green Project Management*, Boca Raton, FL: CRC Press.

Maltzman, R. and Shirley, D. (2013), Project manager as a pivot point for implementing sustainability in an enterprise, in Silvius A.J.G. and Tharp, J. (eds), *Sustainability Integration for Effective Project Management*, Hershey, PA: IGI Global Publishing.

Marcelino-Sádaba, S., Pérez-Ezcurdia, A. and González-Jaen, L.F. (2015), Using project management as a way to sustainability. From a comprehensive review to a framework definition, *Journal of Cleaner Production*, 99, pp. 1–16.

Martens, M.L. and Carvalho, M.M. (2016) Sustainability and success variables in the project management context: an expert panel, *Project Management Journal*, 47(6), pp. 24–43.

Martens, M.L. and Carvalho, M.M. (2017) Key factors of sustainability in project management context: A survey exploring the project managers' perspective, *International Journal of Project Management*, 35(6), pp. 1084–102.

Martens, P. (2006) Sustainability: Science or fiction?, in *Sustainability: Science, Practice, & Policy*, 2(1), pp. 1–5.

McDonough, W. and Braungart, M. (2002) *Cradle To Cradle: Remaking the way we make things*. New York: North Point Press.

Molenaar, K.R. and Sobin, N. (2010) A synthesis of best-value procurement practices for sustainable design-build projects in the public sector, *Journal of Green Building*, 5(4), pp. 148–57.

Office of Government Commerce (2009) *Managing Successful Projects with PRINCE2*, Norwich, UK.

Office of Government Commerce (2010) *Management of Risk: Guidance for Practitioners*, Norwich, UK.

Pade, C., Mallinson, B. and Sewry, D. (2008) An elaboration of critical success factors for rural ICT project sustainability in developing countries: Exploring the Dwesa case, *The Journal of Information Technology Case and Application*, 10(4), pp. 32–55.

Perrini, F. and Tencati, A. (2006) Sustainability and stakeholder management: The need for new corporate performance evaluation and reporting systems, *Business Strategy and the Environment*, 15(5), pp. 286–308.

Project Management Institute (2013) *A Guide to Project Management Body of Knowledge (PMBOK Guide)*, fifth edition, Newtown Square, PA): Project Management Institute.

Robinson, J. (2004) Squaring the circle: On the very idea of sustainable development, *Ecological Economics*, 48(4), pp. 369–84.

Schoper, Y.-G., Wald, A. and Ingason, H.T. (2018) Projectification in Western economies: A comparative study of Germany, Norway and Iceland, *International Journal of Project Management*, 36(1), pp. 71–82.

Shenhar, A.J., Dvir, D., Levy, O. and Maltz, A.C. (2001) Project success: A multidimensional strategic concept, *Long Range Planning*, 34(6), pp. 699–725.

Silvius, A.J.G. (2015) Sustainability evaluation of IT/IS projects, *International Journal of Green Computing*, 6(2), pp. 1–15.

Silvius, A.J.G. (2016) Integrating sustainability into project risk management, in Bodea, S., Purnus, A., Huemann, M and Hajdu, M. (eds) *Managing Project Risks for Competitive Advantage in Changing Business Environments*, Riga, Latvia: IGI Global, University of Latvia.

Silvius, A.J.G. (2017) Sustainability as a new school of thought in project management, *Journal of Cleaner Production*, 166, pp. 1479–93.

Silvius, A.J.G., Kampinga, M., Paniagua, S. and Mooi, H. (2017) Considering sustainability in project management decision making; An investigation using Q-methodology, *International Journal of Project Management*, 35(6), pp. 1133–50.

Silvius, A.J.G. and Schipper, R. (2014) Sustainability in project management: A literature review and impact analysis, *Social Business*, 4(1), pp. 63–96.

Silvius, A.J.G. and Schipper, R. (2015) Developing a maturity model for assessing sustainable project management, *Journal of Modern Project Management*, 3(1), pp. 16–27.

Silvius, A.J.G. and Schipper, R. (2016) Exploring the relationship between sustainability and project success – Conceptual model and expected relationships, *International Journal of Information Systems and Project Management*, 4(3), pp. 5–22.

Silvius, A.J.G., Schipper, R., Planko, J., Brink, J. van der and Köhler, A. (2012) *Sustainability in Project Management*, Farnham, UK: Gower Publishing.

Silvius A.J.G. and Tharp, J. (eds) (2013) *Sustainability Integration for Effective Project Management*, Hershey, PA: IGI Global Publishing.

Tam, G. (2010) The program management process with sustainability considerations, *Journal of Project, Program & Portfolio Management*, 1(1), pp. 17–27.

Taylor, T. (2010) *Sustainability Interventions – for Managers of Projects and Programmes*, Salford, UK: The Higher Education Academy – Centre for Education in the Built Environment.

Tharp, J. (2013) Sustainability in project management: Practical applications, in Silvius A.J.G. and Tharp, J. (eds), *Sustainability Integration for Effective Project Management*, Hershey, PA: IGI Global Publishing.

Turner, J.R. (2010) Responsibilities for sustainable development in project and program management, in Knoepfel, H. (ed.), *Survival and Sustainability as Challenges for Projects*, Zurich: International Project Management Association.

Turner, J.R. (2014) *Handbook of Project Management*, 5th edition, Farnham, UK: Gower Publishing.

Turner, J.R. and Müller, R. (2003) On the nature of the project as a temporary organization, in *International Journal of Project Management*, 21(3), pp. 1–8.

van den Brink, J. (2013) How positive psychology can support sustainable project management, in Silvius, A.J.G. and Tharp, J. (eds), *Sustainability Integration for Effective Project Management*, Hershey, PA: IGI Global Publishing.

Weninger, C. and Huemann, M. (2013) Project initiation: Investment analysis for sustainable development, in Silvius, A.J.G. and Tharp, J. (eds), *Sustainability Integration for Effective Project Management*, Hershey, PA: IGI Global Publishing.

Winnall, J.-L. (2013), Social sustainability to social benefit: Creating positive outcomes through a social risk, in Silvius, A.J.G. and Tharp, J. (eds), *Sustainability Integration for Effective Project Management*, Hershey, PA: IGI Global Publishing.

World Commission on Environment and Development (1987) *Our Common Future*, Oxford: Oxford University Press.

5

TOWARDS SMART FREIGHT LEADERSHIP

Sophie Punte and Floor Bollee

Introduction

Freight transport forms the backbone of today's global economy. It involves the transportation of goods by vehicles or vessels (for modes of transport and key players, see Box 5.1). This sector is facing three major challenges: explosive growth, inefficiencies and rising costs, and high fuel consumption and environmental impacts.

This chapter focuses on the third issue, that is, high fuel consumption and environmental impacts. However, the three major issues are highly interconnected. Therefore, discussing environmental problems in freight transport will inevitably require the consideration of growth in demand and inefficiencies, as both lead to higher energy use and environmental impact.

BOX 5.1 MODES OF TRANSPORT AND KEY PLAYERS

Freight includes inland (rail, road, and inland waterways), maritime, air and pipeline transportation with transhipment centres in between the various modes of transport. Logistics is the management of freight movements and related services between pick-up and delivery to the customer. Key players in the global logistics supply chain are cargo owners (also referred to as shippers), logistics service providers (LSPs), freight forwarders, carriers and customers/receivers. They interact with governments, civil society and other industry players.

Freight and logistics account for about 7–8% of carbon dioxide equivalent (CO_2e) or greenhouse gas (GHG) emissions worldwide. Road freight constituted 8% of international trade in 2015; it, however, generated 40% of CO_2e emissions. When considering all freight, international and domestic, road holds 18% of freight tonne-km but about 67% of CO_2 emissions.

Smart freight

When discussing how to address these environmental issues, smart or green freight comes in. It represents the efforts to transform the freight and logistics sector with a view to reducing GHG emissions and air pollution by improving its fuel efficiency across the global logistics supply chain without affecting the sector's vital economic functions.

In practice, the terms smart freight, green freight and sustainable freight/logistics are used interchangeably.

Smart freight solutions

The freight sector is complex and fragmented. Therefore, the problem must be tackled from different angles. The freight sector is driven by the interaction of the transport system (i.e. infrastructure and policies), freight equipment (i.e. vehicles and vessels) and freight movement (e.g. shipper/customer, LSPs and carrier logistics decisions).

Ideally, the freight sector is supported by:

1   A quality transport system: quality infrastructure and connectivity within and between modes with supporting policy and regulatory frameworks.
2   An optimized freight movement: sector-wide adoption of logistics solutions that maximize load factors and optimize the routing and scheduling of freight flows.
3   Efficient vehicles and vessels: sector-wide adoption of technologies and strategies that offer greater efficiencies and emission reductions.

Smart freight solutions are different at the urban, national and global levels and for different modes; what works best may also vary between countries and regions. Shippers/cargo owners, LSPs and carriers that cover all three levels are likely to meet the most ambitious corporate emission reduction targets.

Existing efforts

Even though there seem to be obvious economic incentives to apply the principles of smart freight, to date there has been little change at a global scale. However, efforts by governments and industry, including research programmes and technological innovation, have given a greater momentum to considering smart freight. Today, there is a growing number of partnerships, such as the International Transport Forum (ITF), the Sustainable Mobility for All (Sum4All) and the Partnership on Sustainable Low Carbon Transport (SLoCaT) that involve green freight as part of a broader range of transport and climate issues.

Multinational firms and their brands undergo a growing pressure of shareholders, customers and governments to reduce their carbon footprint. This market pressure cascades down the global supply chain to multinational carriers and logistics providers and national/local carriers. All of them wish to join the drive to reduce costs, comply with current and forthcoming legislation and retain their 'licence to operate'.

Green freight programmes respond to these business needs. However, such programs do not exist in all countries around the world while they do not have full sector coverage. This holds especially true for road freight, where even the more successful programs like SmartWay cover only 25% of road freight tonne-km (many small carriers are not included as they are difficult to reach).

For this reason, the Climate and Clean Air Coalition (CCAC), hosted by the United Nations, launched the Global Green Freight Action Plan in 2015 to bring governments, private sector, civil society and other actors together to align and enhance existing green freight efforts, develop and support new green freight programmes and to incorporate Black Carbon and air pollutant reductions into green freight programmes alongside GHG emissions.

Smarter freight leadership

To improve efficiency and reduce the environmental impacts from freight transport, the world needs leaders.

A smart freight leader is an organization that demonstrates leadership at three levels: (1) it leads by example, showing commitment to reducing its emissions; (2) it leads firms within its own logistics chains into optimizing these chains; and (3) inside and outside its own logistics supply chain, it leads its stakeholders to work together, standardize monitoring methods and promote the sharing of best practices in the field (Defee, 2007).

However, studies show that only a fraction of cargo owners takes up the role of Smart Freight Leader. A study on marine logistics found that about 40% of cargo owners measure their supply chain emissions and only 6% implement explicit carbon reduction initiatives (McKinnon, 2014).

The top management of logistics service providers (LSPs) tends to consider a green reputation as of secondary importance. Both cargo owners and LSPs consider carrier sustainability as one of the least important selection criteria (Lai et al, 2011; van den Berg and de Langen, 2014).

There is an urgent need for many more business leaders who share and propagate a common vision on how to reach an environmentally sustainable freight and logistics sector.

Smart Freight vision

Our vision is 'Smart Freight, which equates to a journey towards efficient and environmentally sustainable global freight and logistics sector. SFC developed the Smart Freight Leadership Framework to create a common approach for what is needed for government, the private sector and civil society to deliver the vision and achieve the desired transformational change and emission reductions at scale (Smart Freight Centre, 2016a).

As Smart Freight Leaders, businesses take action across their global logistics supply chain to reduce emissions within the 1.5–2°C scenario while creating value for business and society. At centre stage are cargo owners and carriers, with logistics services providers and freight forwarders taking either role. They are supported by the broader private sector, government and civil society. The role of consumers must be considered as end-costumer of many logistics chains.

This vision needs to consider (1) a business perspective, (2) collective leadership and collaboration and (3) linking freight and logistics to climate and sustainable development goals. This is explained as follows.

A business perspective

Freight and logistics are predominantly undertaken by commercial private sector businesses, such as shippers, logistics service providers and carriers. These companies need a clear business case that ensures a satisfactory profit margin. On the other hand, it is recognized that society is involved with its social requirements that companies should meet. It is a matter of shared value.

Core business drivers are the potential to enhance company performance and competitiveness through reduced costs, an improved customer service and less exposure to climate-related risks. Customers, governments and the public will recognize a company's leadership in responsible business through labels, awards or publicity. Further steps may be a matter of an increasing compliance with a future carbon-free world based on new technologies and smarter logistics. Then, there will emerge opportunities to participate in community-based discussions on how to reach a sustainable economy and influence policies in this area.

Amidst all of this, it is important to understand why businesses are slow in reducing emissions. One barrier is a lack of standard methodologies to calculate their carbon footprint and set emission reduction targets. Also, to mention a few more, policy and legislation can be rather unpredictable, fail to create a level playing field or don't work for business.

Collective leadership and collaboration

Freight transcends individual countries and companies. Therefore, many businesses need to broaden their policy perspective to unlock the enormous potential for the freight sector to improve efficiency and reduce emissions. Here, coordinated efforts within industry in partnership with government and civil society are key. We need leadership from all three of them.

Companies can show leadership by integrating smart freight into their business strategies and operational decisions relating to freight across their logistics supply chains. Other private sector players, such as industry associations, technology suppliers and service providers, could contribute, for instance, to target setting, providing input into government policies, investing in new technology and collaborating with other leaders to mobilize industry-wide action.

Governments can demonstrate leadership by providing a stable environment for deploying smart freight solutions and innovation. They can do this by, for example, adding freight to carbon pricing policies, developing national green freight programmes and investing, together with the private sector, in infrastructure that facilitates smart freight.

Civil society can play a crucial role by putting smart freight, climate change and clean air on the agenda. Non-profit organizations, universities and research institutes, development agencies and funders should collectively embrace smart freight as a core issue. Unfortunately, many civil society organizations overlook freight issues. To illustrate this, freight represents about 45% of transport carbon emissions yet receives only 11% of transport climate financing (Gota, 2016).

Often, consumers are not aware of the great distances the goods they buy have travelled before being available in the nearby stores. And yet, consumers are at the root of the demand for the movement of goods. In recent years, e-commerce has expanded enormously, alongside a booming freight and logistics. Therefore, to tackle freight inefficiencies and associated emissions at the root, leadership and concerted action must involve consumers – directly or indirectly. Part of the puzzle is discouraging the instant gratification of needs through price increases that explicitly relate to transportation costs.

Linking freight to climate and sustainable development goals

Freight and logistics do not exist on their own but are very much a 'servant' to other sectors, such as manufacturing, retail, construction and others. For freight and logistics to secure a real spot on the global climate and sustainable development agenda, a link must be made to broader existing efforts that span all main sectors. Two broader international efforts stand out.

The first of these is the Paris Climate Agreement that resulted from the 2015 United Nations Climate Change Conference and sets out a global action plan to keep global warming well below 2°C. Positioning freight and logistics within ongoing UNFCCC conferences and in the implementation of national determined contributions (NDCs) is a matter of concern: while 43% of the NDCs explicitly refer to passenger transport, only 13% mention freight transport explicitly (Partnership on Sustainable, Low Carbon Transport, SLoCaT (2015).

The second framework comprise the UN's Sustainable Development Goals (SDGs); it is a call to end poverty, protect the planet and ensure that all people enjoy peace and prosperity.[1] The 17 goals are interconnected: some have a more or less direct relevance to freight, in particular: affordable and clean energy (goal 7), industry, innovation and infrastructure (goal 8), sustainable cities and communities (goal 11) and climate action (goal 13). The more we are able to link smart freight to relevant SDGs, the easier it is to gain support from decision makers from outside the freight and logistics sector in considering the environmental effects of freight.

The Journey to Smart Freight

Smart Freight Leadership forms the basis for the journey to an efficient and environmentally sustainable global freight and logistics sector.

A common approach is needed for government, the private sector and civil society to deliver the vision and achieve the desired transformational change and emission reductions at scale. Beyond creating a common approach, this framework can be used to spur and guide action to accelerate the uptake of smart freight solutions by industry.

Existing leadership practices are grouped under five behaviours to provide an 'umbrella' for multinationals to deliver carbon emission reductions and cost savings at scale. Smart Freight Leaders aim to

- Calculate and report credible logistics emissions
- Set ambitious targets and KPIs to drive emission reductions
- Take apposite business decisions to approve performance.
- Collaborate with logistics partners and leaders across the logistics supply chain.
- Advocate a long-term strategy and public policy towards decarbonization.

Calculate and report credible logistics emissions

What you don't measure, you can't manage. This also applies to emissions resulting from freight movement and logistics activities. Calculation and reporting of credible logistics emissions should be high on the corporate agenda, especially of companies with high logistics supply chain emissions relative to total corporate emissions.

Challenges

Multinationals operating along global logistics supply chains require a standard multi-modal methodology for emissions calculation and third-party assurance. There are several barriers to measuring emissions, in particular, a universal calculation method and a lack of reliable data. Therefore, we focus primarily on these barriers and what is needed to deal with them.

Lack of universal method for calculating logistics emissions

Companies are increasingly expected to report and systematically reduce their GHG emissions. Tracking GHG emissions can be cumbersome because of the various methodologies and reporting formats that different customers, countries and programmes wish to see used. To create and implement a universal method of calculating logistics emissions, Smart Freight Centre established the Global Logistics Emissions Council (GLEC), which consists of a group of companies, industry associations and programmes backed by leading experts, governments and other stakeholders. The GLEC released the GLEC Framework for Logistics Emissions Methodologies in 2016 to make carbon accounting work for business (Smart Freight Centre, 2016b). For the first time, emissions can be calculated consistently at a global level across road, rail, air, sea, inland waterways and transhipment centres.

Companies that provide or purchase logistics services can use this methodology to:

- Use greenhouse gas emissions as a metric for sustainable freight transportation decisions
- Identify solutions that improve efficiency and reduce emissions and costs
- Track progress towards your climate goals
- Inform customers of emissions reductions achieved
- Stay ahead of regulatory requirements.

An updated GLEC Framework (to be released in 2018), will include details about air, inland waterways and transhipment centres. A separate module to calculate Black Carbon emissions was published in 2017.

A lack of data reduces the reliability of reported carbon footprints and the decisions based on these

A related problem is the availability of reliable data. As cargo owners and LSPs make use of the same pool of millions of carriers, it is not feasible or practical to collect their data separately. Green freight programmes serve as central databases for the collection, analysis and benchmarking/reporting of data of these shared carriers. BSR's Clean Cargo Working Group (see Box 5.2) and SmartWay have the most advanced and complete databases covering marine container freight and road freight in the US and Canada respectively.

Unfortunately, these programmes do not yet span the globe; moreover, only a few have a sufficient coverage of carriers to allow for reliable benchmarking and reporting. This deficiency can be explained by the fact that carriers are wary of sharing detailed fuel and emissions data as that would give shippers a critical insight into their fuel costs and profit margins. Shippers could use this information to negotiate lower contract prices. After all, shippers and carriers have a commercial relationship.

BOX 5.2 BSR CLEAN CARGO WORKING GROUP[2]

The Clean Cargo Working Group (CCWG) was founded by BSR as a global, business-to-business initiative. It is dedicated to improving the environmental performance of marine container transport through measurement, reporting, evaluation and best practice sharing. Today, CCWG tools represent the industry standard for measuring and reporting ocean carriers' environmental performance on CO_2 emissions and other environmental impacts. CCWG has over 50 members and carriers reporting represent over 85% of ocean container cargo. Every year CCWG produces global trade lane emissions factors which have shown an industry reduction in CO_2 emissions of 29% per TEU-km since 2009. CCWG creates practical tools for measuring, evaluating and reporting the environmental impacts of global goods transportation, helping ocean freight carriers track and benchmark their performance and readily report to customers in a standard format, and cargo owners (shippers) review and compare carriers' environmental performance when reporting and making informed buying decisions.

Limited capacity and demand for emissions calculation and reporting

A survey of cargo owners found that several other barriers exist to measuring emissions (McKinnon, 2014). The most significant barriers relate to capacity (resources, time, expertise, knowing how to measure and report) and demand from customers and governments.

Many small and medium-sized enterprises (SMEs) in the logistics sector confront a knowledge and capacity gap. However, even the largest logistics service providers, who outsource much of their transport to specialist carriers, struggle with acquiring reliable, company-specific data in a format that fulfils their customers' carbon accounting needs.

Proposed solutions

Proposed solutions all evolve around promoting emissions accounting and emission reduction: adopting the GLEC Framework, developing data interfaces between shippers and carriers and creating incentives and support mechanisms for calculation and reporting.

Adoption of the GLEC Framework for Logistics Emissions Methodologies

The next step towards unifying carbon accounting practices in the logistics sector is a matter of promoting the acceptance and use of the GLEC Framework by industry, governments and other players. Part of this is the adoption of the GLEC Framework by companies, especially multinational shippers, LSPs and carriers that participate in industry-backed green freight programmes, industry associations relevant to the freight sector and industry networks with logistics elements. GLEC member companies are not only champions of the GLEC Framework but can also promote its adoption by reaching out to other companies by means of green freight programmes involving industry associations and business networks in which they participate. Leading multinationals, including HP Inc, Intel, DB Schenker, Deutsche Bahn DHL Group, Geodis, Kuehne & Nagel and SNCF, are committed to adopting the GLEC Framework.[3] Second, more companies will apply the GLEC Framework if it is integrated in calculation tools used by companies to calculate emissions, which can be a company's in-house tool or tools available commercially such as NTM, EcoTransIT and TK Blue. Tools can be aligned with the GLEC Framework by, for example, creating a 'GLEC default' setting.

Green freight programmes can be more effective in enabling their company members to identify and deliver on emission reduction opportunities by adopting the GLEC Framework as it enhances comparability and usefulness of emissions calculations.

Companies are more likely to join programmes that are effective and consistent with those for other modes/regions. Harmonizing logistics emissions accounting through the GLEC Framework was already a key action in the Global Green Freight Action Plan (Climate and Clean Air Coalition 2015). Lean & Green Award members developing reduction plans follow the GLEC Framework to make data more comparable. The Carbon Disclosure Project (CDP) lists the GLEC Framework in its 2017 Climate Change Reporting Guidance.[4]

Standards for freight and logistics emissions issued by governments or other institutions will gain in credibility if they operate in line with the industry-backed GLEC Framework. To maximize the mutual alignment among applicable updates of existing standards, such as EN 16258, Article L.1431–3 of the French Transport Code or 'French Grenelle', the GLEC Framework could be very helpful. In turn, the credibility and use of the GLEC Framework will be enhanced if it is used as the basis for a new ISO standards and embedded in related ISO standards.

Develop data interfaces between shippers and carriers taking commercial sensitivity of data into account

The quality of calculated emissions depends on the availability of quality data. Here, there is a data gap whose closing has to take priority, particularly in the area of road and air freight.

First, there is an immediate need to develop 'consumption factors' (CO_2e or fuel per tonne-km) for different modes and regions, which will improve the accuracy of a company's calculated footprint. This can be done together with institutions that have relevant data: for example, using the data collected through green freight programmes like ObjectifCO$_2$ and the Logistics Emissions Reduction Scheme, it seems possible to derive consumption factors for France and the UK respectively that are superior to the current default factors. Another option is to conduct surveys, for example of trucks driving along busy freight corridors in Europe, Africa or Asia, to derive corridor consumption factors, similar to the shipping lane factors available through Clean Cargo Working Group for ocean container freight.

Second, the data exchange between shippers and carriers needs to be improved by a significant investment without compromising a carrier's commercial position. The pros and cons of the following options should be analysed to determine the best option for different countries and regions:

- Carriers report primary data (fuel use, distance travelled. etc.) to online tools/databases embedded in green freight programmes to calculate fuel or GHG consumption factors centrally.
- Carriers make their own calculations and report total emissions and consumption factors to green freight programmes. Green freight programmes subsequently compile reports like SmartWay.
- Carriers make their own calculations and report their total emissions and consumption factors using a standard template accepted by many carriers and shippers. Shippers can then develop an interface to integrate the reports from all their subcontracted carriers.
- Carriers make use of interface application(s) for data exchange with customers. Data can come from carriers directly or be pulled from emission calculation tools or databases.

The staggering ICT innovation in transport management systems in the logistics sector could facilitate more reliable fuel and GHG emissions data.

Develop incentives and support mechanisms for emissions calculation and reporting

Businesses acknowledge that without the inclusion of emissions calculation and reporting requirements in carrier contracts or a legal requirement to report emissions, proper emission figures remain a mere voluntary undertaking. However, the various green freight programmes or similar initiatives such as the Carbon Disclosure Project that require measurement, reporting and verification (MRV) of simulated data are obvious incentives to produce the required data and in return give companies a distinguishing label, award or another form of recognition. See Box 5.3.

'Labels' are a frequently mentioned incentive mechanism; they show three options: (1) as a separate GHG figure for logistics services/activities, (2) inclusion in a product life cycle GHG figure, and (3) embedded in label schemes of green freight programmes that combine GHG data/reporting with other programme requirements.

Collaborative green freight programmes seem to be a logical path to spread a wide-reaching, well-designed B2B label in the logistics sector, but such a label

---

BOX 5.3 LOGISTICS EMISSIONS ACCOUNTING AND REDUCTION NETWORK (LEARN)

The project 'Logistics Emissions Accounting and Reduction Network' (LEARN) empowers business to reduce their carbon footprint across their global logistics supply chain through improved emissions accounting. Logistics emissions measurement, reporting and verification (MRV) is improved and accelerated in four ways:

1   Provide support to companies through guidance, training and education, and develop a blueprint for a label.
2   Test and validate with companies the practical applicability of emissions MRV and requirements for a label in complex multi-modal logistics settings.
3   Promote and facilitate supportive policy and research.
4   Develop and involve a LEARN multi-stakeholder network to maximize business uptake of carbon accounting and carbon reduction.

LEARN builds on existing initiatives and networks and the GLEC Framework. LEARN is funded by the European Union's Horizon 2020 research and innovation programme.

---

needs to transparently underpinned and harmonized between programmes and serving the purposes of various user groups.

Set ambitious targets and KPIs to drive emission reductions

Defining key performance indicators (KPIs) and setting targets are used to ensure that the global freight and logistics sector takes its share in keeping global atmospheric CO_2 concentrations below the level that corresponds with a 1.5–2.0°C temperature rise compared to pre-industrial times. KPIs and target setting are applicable to companies, countries and sectors. Companies that use emission-related KPIs and set targets have the benefit of knowing what they are aiming for and thus can avoid failing to deliver figures that cannot be accumulated.

Existing challenges

To date, efforts have been made to define KPIs and set targets. However, companies often set targets that are inconsistent; national or sector targets from which to derive business targets are mostly lacking; and green freight programmes lack a unified approach, which may give slow starters an undesirable advantage.

KPIs and target-setting by companies

Senior managers can show they are serious about decarbonization by setting ambitious emission reduction targets for the company's global logistics supply chain. These targets should be supported by strong corporate policies favouring low carbon freight and logistics and KPIs at every level of the organization. Only when logistics and operations directors are being held accountable for meeting these targets will they take them seriously. Then, all their business decisions will be influenced by these targets, such as the selection of subcontractors, transport modes, routes as well as locating distribution centres, investing in vehicles and equipment.

TABLE 5.1 Examples of business emission reduction targets reproduced from McKinnon & Piecyk (2012)43 and DP/DHL (Connekt, 2016)

Company	Normalizer (KPI)	Time period	Target as % reduction in carbon intensity
Deutsche Post/ DHL	Every letter and parcel delivered, every tonne of cargo transported and every square metre of ware-house space	2007–2050	50% by 2025 Zero emissions by 2050
DB Schenker	Tonne-km	2006–2020	At least 20%
UPS	UPS Transportation Index	2010–2017	5%
UPS Airlines – Global	Pounds of CO_2 emitted for every ton of capacity transported per nautical mile	2005–2020	20%
FedEx	Available ton miles (ATM, tons of carrying capacity multiplied by miles travelled)	2005–2020	20%
TNT (Mail and Express)	Not specified: only refers to CO_2 efficiency		45%
Maersk Line	Not specified: only refers to CO_2 efficiency	2007–2020	25%
NYK	Unit of transportation from vessels	2006–2013	10%

Companies can set targets in two ways. Absolute emission reduction targets relate to the total quantity of greenhouse gas emissions being emitted. Intensity reduction targets are expressed as emission reductions per unit of an operational or economic output.

Table 5.1 gives examples of business targets and shows that companies express their targets differently, leading to comparison of apples and oranges. Moreover, as companies set emission intensity reduction targets, whether these will translate into real emission reductions is highly dependent on growth in freight volumes or tonne-km.

Green freight programmes and initiatives have different target requirements

There are also several green freight programmes that require companies to set targets and deliver tangible results, but requirements differ among them (Smart Freight Centre (2015):

- Lean and Green (Netherlands and other European countries). Members commit to a minimum target of 20% CO_2 reduction per tonne-km (or comparable unit) over a maximum 5-year period covering at least 50% of their operations.
- Logistics Emissions Reduction Scheme (UK). Programme members are collectively committed to 8% reduction in the intensity of CO_2e emissions of their 77,000 trucks and vans by 2015, compared to a 2010 baseline.
- Objectif CO_2 (France): Member carriers, currently covering around 20% of commercial road freight vehicles in France, must set a CO_2 reduction target against two performance indicators: gCO_2/km and $gCO_2/tonne$-km over three years.

These programmes make use of percentage reduction targets – expressed as either absolute emissions or emissions intensity. This creates an unfair advantage for companies that have made little effort in the past to reduce their emissions, that is, it is easier for them to improve when they join green freight programmes.

One effort that has gained traction is the Science-Based Target Initiative (SBTI) of the World Resources Institute (WRI), World Wildlife Fund (WWF), UN Global Compact and the Carbon Disclosure Project (CDP).[5] The basis for setting targets is the emissions budget facilitating a global warming increase of <2°C by 2100. By early 2018, over 300 companies had committed to setting science-based targets following clearly defined steps. However, logistics only need to be included under 'scope 3' if it covers a significant portion of a company's overall emissions (Ministry of Transport, 2016).

In 2009, the European Commission enacted a climate and energy package as a set of binding legislation, which sets clear targets for 2020, known as the 20–20–20 targets; however, no targets for freight and logistics as a sector were specified.[6]

Nonetheless, The Netherlands' Top Sector Logistics, in partnership with NGO Connekt, promotes an 80% emission reduction target in a business-like manner,

'Factor 6': in 2050, the logistics sector intends to work six times more efficiently than it did in 1990.[7]

Sectoral targets were set by, for instance, the airline sector, which in 2012 emitted 677 million tons CO_2, representing 2% of global emissions. As commercial cargo represents no more than approximately 0.5% of the total volume (although about one-third of total value) transported by air, the International Air Trade Association (IATA, representing 240 airlines covering 84% of global air travel) adopted the following targets to mitigate CO_2 emissions from all air transport:[8]

- An average improvement in fuel efficiency of 1.5% per year from 2009 to 2020
- arbon-neutral growth: a cap on net aviation CO_2 emissions as from 2020
- reduction in net aviation CO_2 emissions of 50% by 2050, relative to 2005 levels.

Proposed solutions

Leadership and collaboration are needed to make target setting more consistent, transparent and fair.

Develop a common set of KPIs and methodology for setting business emission reduction targets

A common set of KPIs for freight efficiency and GHG emissions will help company managements improve performance. Within the freight and logistics sector, agreement should be reached on a common, global methodology for setting targets and how these must be expressed. Only then can emission targets and actual reductions be added together so as to assess whether companies achieve sizeable emission reductions across a country or mode of transport. This should build on existing efforts, such as the SBTI, green freight programmes, and modal targets.

Moreover, KPIs must be properly defined. For example, GHG per tonne-km may be too generic a measure to drive change. A more effective KPI could be the number of trucks withdrawn because this measure would encourage an increase in truck filling rates (including an effective use of return trips) and shifting freight to other, more energy-efficient, modes of transport. Such an improvement will only take place if based on appropriate corporate policies; in this case, a company must be prepared to make the most of its transport capacity by involving the transport needs of fellow companies.

Improve target setting requirements under Green Freight Programs and initiatives

To make target setting more fair and transparent, programs should align target setting and make use of benchmarks for CO_2 per tonne-km targets. The US SmartWay program, while not mandating target setting, categorizes companies into four groups (for each truck/service type) based on their ton CO_2/tonne-km performance across the fleet, thus allowing their customers and other stakeholders to compare companies on an equal basis.

As for SBTI, it is advisable that companies committing to target setting disclose what percentage of their business's total carbon footprint comprises freight and logistics. This way it can be determined if these are significant and should be included in company targets under this initiative.

Improve methodology alignment between countries, modes and companies

Business targets for freight and logistics ideally should be put into context of gobal, national and sectoral targets -after all, all countries and sectors must together work towards a low carbon world.

To make this happen, ideally global targets should be set on the emission ceiling that prevents global warming surpassing an increase of 1.5–2 degrees C as confirmed by the Paris Climate Agreement. Using the same or consistent methodology, national and sectoral targets should be derived from that, including emission redution targets for the global freight and logistics sector, and modes within these. This in turn will allow business to check that their own targets are in line with targets of sectors and countries in which they operate.

As mentioned before, it may be harder and less cost-effective to cut the carbon intensity of transport than of other sectors. An effort to set such a target for the global freight sector in its totality and/or by mode, should thus take freight characteristics into account, such as:[9]

- Relative cost of decarbonizing freight transport in comparison to other sectors, especially as available evidence suggests these are higher for transport including freight.
- Interdependence between freight transport and other sectors, acknowledging that the demand for freight is highly dependent on developments in other sectors.
- Technological and sectoral developments, including the full range of decarbonization options available, particularly on the softer behavioural/ operational side. For example, for transport broadly, electric vehicles could trigger a revolution in this sector, and if combined with a shift from conventional fuels to renewable fuels in electricity generation this would have massive impacts on freight CO_2 emission levels.

Take apposite business decisions to improve performance

The rubber hits the road with an Action Plan containing concrete measures to reduce the carbon footprint. Action Plans or roadmaps can be developed by countries (following their Nationally Determined Contributions or NDCs) and by companies.

Existing challenges

Solutions are known, yet we seldom see their uptake at scale in the logistics supply chain of multinational shippers, LSPs and carriers. Reasons are insufficient guidance and examples, dependency on subcontracted carriers who face barriers, and support efforts not reaching businesses.

Insufficient guidance and examples to develop action plans for logistics supply chains

Guidance on how to develop an effective action plan to curb logistics emissions is hard to find. Examples of actions plans are also scarce because companies usually don't share these publicly. Despite case studies and examples of smart freight solutions on various websites and reports, businesses often complain that these are too generic and not specific for their situation, which is not surprising given the vast differences within the sector.

Shippers depend on subcontracted carriers who face barriers to implementing actions

Multinational shippers, LSPs and carriers that want to reduce their carbon footprint are to a large degree dependent on what their subcontracted carriers do. While this is the case for all modes, this especially applies to road freight due to the many layers of subcontracting and carrier fragmentation – for example, small contractors using old vans for urban deliveries. Barriers exist, both internal and external. Internal barriers relate to obstacles within the company boundaries and that are mostly within management's control. There are four categories of internal barriers: re-active management, management systems, finance, and awareness and capacity.

External barriers relate to outside factors and players over which carriers have little or no influence. The four categories of external barriers are: market forces, institutions and policies, partners and programmes, and technologies and measures.

All these barriers are interconnected and often re-enforce each other. To gain a holistic understanding of existing barriers we must realize that the global freight sector is first and foremost a commercial sector. Market forces, an oversupply of carriers and high fragmentation have created a cutthroat competitive sector that puts carriers under an enormous pressure to survive. This in turn contributes to reactive management within carriers, their ability to finance technologies and measures and invest in management systems and internal capacity building. The other external barriers further weaken carriers' ability to become more fuel efficient.

Fragmented government, research and other support efforts do not reach businesses

Government support aimed to help business take action can take many forms, such as subsidy schemes and information about technologies, but relatively little effort is made to ensure that businesses know about this. Furthermore, while pilot projects can be incredibly useful to demonstrate the potential of technologies and other solutions, efforts often stall at the pilot phase, at a company's boundary or in developed countries for companies that operate worldwide.

Similarly, considerableresearch is conducted on efficiency in logistics but businesses are not always heard in selecting research topics and focus, and the gap between research results and practical application is too wide or results never reach businesses. For example, there are ample examples of actions by companies and several online databases with case studies exist, but these are not easy to locate. Fortunately, the number of initiatives that combine research and business partnerships is on the rise, such as the Dutch Institute for Logistics (Dinalog) and the Centre for Sustainable Road Freight.

Green freight programmes provide support through guidance, technology verification, financing schemes, other support, or a combination of these. The toolboxes of SmartWay, ObjectifCO$_2$ and EcoStars stand out for their practical nature and help companies make more informed purchases. SmartWay's Technology Program tests and verifies emissions reductions and fuel savings for various available technologies, such as aerodynamics, idle reduction technologies and low rolling resistance tires. SmartWay also provides case studies, fact sheets, technical bulletins and educational materials on fuel-efficient technologies, fleet movement and modal shift. ObjectifCO$_2$ developed freight best practices catalogue with 'Action Sheets' for 54 solutions tailored to four different truck types, and these are now being translated from French into English and Spanish. EcoStars developed a fleet efficiency Road Map, to help carriers progress through the scheme's 5-star rating system, that covers actions in six categories. However, governments and associations in most other countries, let alone shippers and carriers, are not aware that this information exists.

As a consequence, the large majority of companies are not reached with the support they need.

Proposed solutions

Guidance and examples to develop action plans, more carrier initiatives and clarity of support options that are available to businesses can help address these challenges.

Provide guidance and collect examples to support companies in developing action plans

Guidance to develop action plans, along with examples, is key. Such guidance can draw from green freight programmes such as ObjectifCO$_2$ and EcoStars, corporate social responsibility (CSR) reports, and even action plans or road maps by cities and countries.

While action plans vary in structure and content, based on existing plans and strategies reviewed, the content is generally composed of 'where are we now?

Where are we going? How do we get there? And what do we need to implement the plan?'

Develop more initiatives dedicated to help carriers overcome barriers

It is important to focus on what barriers can be realistically addressed, in a concerted effort that looks at all identified barriers holistically, while involving carriers, their immediate partner and broader stakeholders. First, while it is tempting to focus on market forces, this is not realistic. A better approach is to focus on barriers over which carriers have more control, such as management systems or awareness and capacity, or that partner NGOs and government agencies can address, such as policies and programmes. Second, addressing one barrier is not going to solve the problem. For example, a combination is needed of clear policies and technology standards that can build trust in technologies available on the market; fleet managers' ability to assess which technologies and suppliers are the best fit; good management systems that generate data for a strong business case to management; and a solution to overcome cash flow challenges to invest.

More initiatives that explicitly target carriers and their specific barriers can make a real impact. Examples of existing initiatives include:

- Green freight programmes built around carriers, such as EcoStars or France's ObjectifCO$_2$
- Initiatives that focus primarily on technologies and measures that carriers can adopt independently, such as Trucking Efficiency in the USa, Low Carbon Vehicle Partnership in the UK and Green Truck Partnership in Australia
- Training courses for drivers and transport managers focused on fuel and operational efficiency, such as SAFED and FORS in the UK and FleetSmart in Canada (now integrated with SmartWay).

Map government, research and support by action type and establish channels to reach businesses

Addressing barriers holistically means that active collaboration and concerted efforts are needed. For example, a green freight programme can provide information on technologies and measures but will be more effective if backed by government and NGOs. Technology providers, industry associations and academic institutions can provide fleet managers training, but this will be more effective if successful fleet managers are involved in training delivery.

Clarity of what support is available to businesses is key, and therefore mapping, consolidating and channelling information to companies is recommended. One company executive put it like this: 'We need someone to provide an "alert function" on broader developments, such as funds, policy developments, studies, initiatives, which would really help strengthen the business case for improvement measures'.

This also means that we should look beyond green freight programmes to reach businesses and look at other programmes and initiatives that target a specific sector or are more mainstream business but with a sustainability angle to them. One example is EPEAT that is managed by the Green Electronics Council and rates electronics products on their environmental performance across the life cycle – thus including transportation. Another is the Dow Jones Sustainability Index.

Collaborate with logistics partners and leaders across the logistics supply chain

Existing challenges

Collaboration is hampered because it is not well understood, neutral players are often missing and the multitude of initiatives is especially confusing to businesses.

Collaboration is diverse and poorly understood with too few published examples

There is ample scope for such collaboration and the potential for cost and environmental savings is huge, as witnessed by available examples. The challenge is that right now only a handful of companies are working together in this way and there are too few published examples. An underlying reason is that collaboration types, principles and process are poorly understood.

Lack of neutral facilitators of collaboration efforts

Individual companies that operate along the same logistics supply chain or have a stake in a common challenge of the freight sector can establish win-win collaborations. To date, however, shippers that are willing to collaborate are too often hindered by practical obstacles. One of these is the availability of neutral and strategic conveners to bring different parties together.

Proliferation of initiatives risking business disengagement

The momentum for climate action generated by the Paris Climate Agreement, combined with a growing realization that freight and logistics is part of that puzzle, is now resulting in new freight initiatives and funders. We need more efforts and funds, but further proliferation must be avoided.

Proposed solutions

What it takes to enter into collaboration does not seem well understood. The many initiatives that have emerged can be quite confusing, while there is a lack of neutral players that can assume a mediating role.

Define principles and process of successful collaboration in logistics with real examples

Successful collaboration needs to be defined and agreed on before putting it into practice. Various stakeholders play a part in it. The following could kick off the discussion on a common framework.

There are three types of collaboration (modified from Forum for the Future) (see also boxes 5.4, 5.5 and 5.6):

1   Within the context of logistics supply chain collaboration, a cargo owner motivates its LSPs and subcontracted carriers to act.

BOX 5.4 WITHIN-LOGISTICS SUPPLY CHAIN COLLABORATION: BACKHAUL FREIGHT BY WALMART AND UNILEVER IN CHINA[11]

The green logistics collaboration between Walmart and Unilever in China dates back to 2009. Unilever delivers its goods to Walmart Distribution Centers (DC), which Walmart transports to its 300+ stores in over 100 cities across China. Both companies had empty trucks on backhaul trips: Unilever when returning from Walmart DCs and Walmart when returning from its stores. The collaboration meant that several trucks returning from Walmart stores picked up Unilever products to take to Walmart DCs. The result: 520 fewer empty truck trips and 58 tons of CO_2 reductions per year, a 10% transport cost reduction for Unilever and 100% on-time pick-ups.

BOX 5.5 CROSS-LOGISTICS SUPPLY CHAIN COLLABORATION: BUNDLING SUPPLIERS' PRODUCTS FOR JUMBO[12]

Retailers like Jumbo face the everyday challenge of getting their products to shops and consumers across The Netherlands using hundreds of carriers. Imagine products from different brands being transported separately to Jumbo's distribution centre, only for trucks to then return empty. There has to be a smarter way. Jumbo has been able to 'bundle' orders from four different product suppliers and make arrangements for freight backhauls. The result: 40% fewer supplies stacked at the distribution centre, 40% fewer deliveries, 30% higher truck load factors, 35% fewer transport kilometres and 35% lower carbon footprint. And of course significant cost savings.

BOX 5.6 MULTI-STAKEHOLDER COLLABORATION: SMARTWAY TRANSPORT PARTNERSHIP[13]

The SmartWay Transport Partnership is the flagship programme of the USA and Canada for improving fuel efficiency and reducing GHG and air pollution from the transportation supply chain industry. The programme was launched in 2004 as a public–private initiative between US EPA and partners including industry stakeholders, environmental groups, the American Trucking Association and BSR. It aims to increase the availability and market penetration of fuel-efficient technologies and strategies that help freight companies save money while reducing adverse environmental impacts. SmartWay is the most extensive and mature green freight programme in operation today. By 2015, over 3,000 members collectively saved $16.8 billion dollars in fuel costs, 120.7 million barrels of oil (the equivalent of taking over 10 million cars off the road for a year), 51.6 million metric tons CO_2, 738,000 tons NO_x and 37,000 tons PM.

---

2   Horizontal collaboration takes place across logistics supply chains, such as between multiple cargo owners who have the same customer(s) or destination market.
3   Multi-stakeholder collaboration on a common challenge or opportunity, involving actors from the logistics sector and government, private sector and/or civil society in areas such as infrastructure projects and technological innovation.

Of these three types, the across logistics supply chain and multi-stakeholder collaborations have the greatest potential to transform the logistics sector.

Give civil society a greater role in collaborative action

Over recent decades, NGOs have undergone an evolution: they started with a focus on charity focus but became more activist in the course of time. Several developed as leaders in advocating particular causes.

The potential role of leading NGOs in facilitating system shifts is only now becoming evident. Too often, only government and businesses were considered to engage in 'public–private partnerships'; however, NGOs deserve a more prominent place in community-focused collaborative action.

Promote alignment and collaboration between initiatives

Policies and actions to curtail GHG emissions in freight and logistics will be more effective if they are aligned and coordinated where possible. Companies will be more interested to join such initiatives if they complement each other and if they are mutually consistent across different modes of transport and markets. An example of such a type of collaboration is the Global Action Agenda (before the Lima Paris Action Agenda), which at present brings together 18 initiatives relevant to transport, with the support from SloCaT/PPMC.[14]

The Alliance to Save Energy is trying to align actions on energy efficiency in the USA and is now going global (Carpenter, 2010[15]). Similar efforts are taking place elsewhere, such as We Mean Business,[16] which brings together leading global business networks on a common climate agenda, the Partnership on Sustainable Low Carbon Transport (SLoCaT)[17] and finally for freight, the Global Green Freight Action Plan (Climate and Clean Air Coalition, 2015) to align existing green freight programmes.

These actions and programmes do not mean much without being transparent about their impact in a similar way which SmartWay quantified its impact up to 2015: over 50 million tonnes of CO_2 and 16.8 billion dollars in fuel savings, equivalent to laying up over 10 million cars for one year.[18]

Engage proactively in public policy development

Government can facilitate and support emissions-mitigating actions by businesses through planning, infrastructure, legislation, regulation, standards and financing. The World Bank's Logistics Performance Index (LPI) gives a clear explanation of the link between good policy and good logistics.

Good public policy making is critical to business' success, and also when it comes to improving logistics efficiency and reducing emissions. Therefore, most companies have joined one or more industry associations to reach out to governments and become actively engaged in public policy making.

Existing challenges

Freight is not on the (climate) agenda of many governments

Several governments have recognized the importance of an efficient and competitive freight and logistics sector for the economy and the environment. For example, the USA has a mix of standards regulating vehicle emissions and technologies; SmartWay is its green freight programme, while there are various other policies and schemes. Similarly, The Netherlands has a mixture of programmes in this field, including the Top Sector Logistics as a public–private endeavour and Lean and Green as a national green freight programme. Several cities developed urban freight plans, including London, Stockholm, New York and Seattle. The Singapore government has announced the introduction of a carbon tax of $Singapore 10–20 per tonne carbon for direct emitters, thus including fossil fuel users. Generated income will be used to finance mitigation measures and innovation efforts under its Climate Action Plan, and has launched a clear price signal to industry to reduce emissions.

However, the above countries are the exception rather than the rule. At government levels, particularly in developing countries, there seem to be a limited awareness of the environmental impacts of freight and logistics. Apart from awareness and capacity to deal with the issue, there could also be another reason. The transport sector is considered to be harder to decarbonize than other sectors, especially in the area of freight (IPPC, 2014). There are exceptions; for example, the Delhi Freight Corridor was built with freight efficiency and less environmental impact in mind; it has the potential to save more than 450 million tons of CO_2 over 30 years (SLoCaT, 2015).

Mixed messages on what business wants from governments

Businesses may support the same objectives as governments but may have different views on how to reach them. This possible discrepancy may lead to a confusingly mixed messaging. Another complicating factor could be that industry associations not only represent leading companies but have to consider all of their members.

A start has been made by We Mean Business as a coalition of organizations working with the world's most influential businesses and investors to accelerate the transition to a low-carbon economy. Its combined 554 company members drafted a list of eight 'policy tasks' in preparation for the UNFCCC summit in December 2015, and that were taken on board in drafting the Paris Climate Agreement.

The UN Secretary-General's High-Level Advisory Group on Sustainable Transport has recommended the following areas for policy development and implementation relevant to sustainable transport, including freight: transport planning, government frameworks, technical capacity of transport planners and implementers, road safety, public engagements and monitoring and evaluation framework including more reliable data (SLoCaT, 2015).

However, when it comes to freight and logistics specified according to countries or markets, there is still a clear gap. In some cases individual companies can influence policies for the better due to their in-depth knowledge of the sector. For instance, Scania has contributed to the Chinese GB1589 standard on dimensions, axle load and masses for motor vehicles, trailers and combination vehicles. This standard will accelerate the transition to a more efficient and safer truck fleet in China.[19]

Fragmented and conflicting government policies

As the various responsibilities for freight and logistics are spread across multiple ministries as well as divided between national and local governments, it is not surprising that government policies and related efforts are often fragmented, overlapping and, in some cases, even conflicting. Several countries, such as Indonesia, continue to subsidize fuel, while at the same time promoting fuel efficiency of vehicles. Alternative fuels are promoted widely, such as natural gas in Thailand but lack of infrastructure to supply the fuel is preventing carriers from using alternative fuels. In China two fuel economy standards were developed for heavy duty vehicles by two different ministries – the Ministry of Transport and the Ministry of Industry and Information Technology – creating uncertainty in the market (Clean Air Asia and World Bank (2013). A comprehensive overview of policies and standards in relation to diesel is available on www.dieselnet.com/standards/

Proposed solutions

Governments are urged to include freight in their national and local plans to curb emissions. The INDCs as part of the Paris Climate Agreement provide a great opportunity to place freight higher on the agenda. There are many NGOs and research institutes with in-depth expertise on transport and climate policies that can provide the required information to do this. UN SDGs and other measures on transport CO_2 emissions.

Next steps involve developing guidance to businesses in expressing their policy needs to governments. Industry-backed initiatives can help ensuring that industry stays ahead of the curve and ensures a proactive and constructive policy influence. Examples include WeMeanBusiness, Climate Finance and green freight programmes.

Furthermore, the development of national and city green freight plans must on the agenda. Freight issues may create bothersome obstacles to policy makers and urban planners in developing sustainable urban freight systems as part of sustainable and liveable cities. To address these challenges, several cities, regions and countries have developed freight plans. A 'freight plan' is a plan with the long-term perspective of establishing a safe, efficient and environmentally sustainable freight system. Existing freight plans not only consider direct environmental issues but also take broader sustainability goals into account.

Cities that are known to have a city freight plan include Berlin, Brussels, Paris, London, Seattle and Stockholm. California and Washington State have regional freight plans. Some countries – including Germany, The Netherlands and the UK – are undertaking 'road mapping' exercises to decarbonize freight through technological and operational measures (Smart Freight Centre, 2017).

The European Commission has released a European Strategy for low emissions mobility, including freight, that focuses on increasing the efficiency of the transport system, speeding up the deployment of low-emission alternative energy for transport and moving towards zero-emission vehicles (European Commission, 2016).

Regional, national and city freight plans can provide the umbrella for policies and initiatives and cohesion between them.

Conclusion

Companies, in particular multinationals, are experienced in developing and implementing strategies to reduce their environmental impact. However, to date, in many cases, such strategies did not consider freight and logistics. Therefore, companies seek guidance in selecting the right solutions and how to mobilize subcontracted carriers and other supply chain partners to take action.

This chapter has shown the relevance and need of highlighting the business case for embracing smart freight solutions.

Companies that participated in the research done have reported that the absence of a global methodology for logistics emissions measurement, reporting and verification across all modes of transport was a serious impediment to demonstrating climate leadership.

For that reason, industry mandated SFC to create a universal method through the Global Logistics Emissions Council (GLEC), resulting in the GLEC Framework for Logistics Emissions Methodologies.[19]

This chapter discussed the various action plans and the progress made in selected countries in tackling the environmental and climate issues relating to freight and logistics. It stresses the need for collective action, alignment and harmonization involving the different modes of transport and regions worldwide. The Smart Freight Leadership Framework, incluing the GLEC Framework, can play a major role in this process.

Notes

1 www.undp.org/content/undp/en/home/sustainable-development-goals.html
2 www.bsr.org/cleancargo
3 www.smartfreightcentre.org/news/first-carbon-accounting-method-launched-for-global
4 www.cdp.net/en
5 www.wemeanbusinesscoalition.org/content/adopt-science-based-emissionsreduction-target
6 http://ec.europa.eu/clima/policies/package/index_en.htm
7 Chinese Standard. GB 1589–2016, Limits of dimensions, axle loads and masses for motor vehicles, trailers and combination vehicles. Available at www.chinesestandard.net
8 www.iata.org/policy/environment/pages/climate-change.aspx
9 Alan McKinnon (2016). Should we set a global carbon reduction target for freight transport? www.alanmckinnon.co.uk/blog/?p=170
10 Walmart China & Unilever China, 'Backhaul Project Green Logistics & Cost Win-Win', ECR AP Conference 2012, www.ecr-all.org/upload/blogfiles/824/Walmart%20China%20%20%20Unilever%20China%20Backhaul%20Project%20final.pdf
11 www.purebusinessboost.nl/uncategorized/grote-fmcg-producenten-werken-met-jumbo-supermarkten-aan-duurzame-groei/
12 www.epa.gov/smartway
13 http://newsroom.unfccc.int/climate-action/global-climate-action-agenda/ and www.slocat.net/ppmc
14 www.ase.org/
15 www.wemeanbusinesscoalition.org/
16 www.wemeanbusinesscoalition.org/
17 www.epa.gov/smartway/
18 See note 7.
19 www.bsr.org/cleancargo

References

Clean Air Asia and World Bank (2013) China green freight policy and institutional analysis report. Available at http://cleanairasia.org/wp-content/uploads/2015/07/China-Green-Freight-Policyand-Institutional-Analysis-Report.pdf

Climate and Clean Air Coalition (2015) Global green freight action plan. Available at www.unep.morg/ccac/Initiatives/ReducingEmissionsFromHeavyDutyDiesel/tabid/1335 73/Default.aspx

Connekt (2016) Factor 6 Magazine. Available at www.connekt.nl/en/nieu ws/factor-6-magazine

Defee, C. (2007) *Supply Chain Leadership*. Dissertation. Knoxville, TN: The University of Tennessee.

European Commission (2016) European strategy for low emissions mobility. Available at http://europa.eu/rapid/press-release_MEMO-16-2497_en.htm

Gota, S. (2016) Freight Transport and Climate Change – Is Freight in Climate Mitigation Agenda? Presentation at Transport Research Board 95th Annual Meeting, January 2016.

Hutch Carpenter (2010) Three types of collaboration that drive innovation. Available at www.cmswire.com/cms/enterprise-20/three-types-of-collaboration-that-driveinnovation 008292.php

IPPC (2014) Fifth Assessment Report (APR). Available at www.ipcc.ch/report/ar5/

Lai, K.H., Lun, V.Y., Wong, C.W. and Cheng, T.C.E. (2011) Green shipping practices in the shipping industry: Conceptualization, adoption, and implications, *Resources, Conservation and Recycling*, 55(6), pp. 631–8.

McKinnon, A. (2014) The possible influence of the shipper on carbon emissions from deep-sea container supply chains: An empirical analysis, *Maritime Economics & Logistics*, 16(1), pp. 1–19.

McKinnon, A. and Piecyk, M. (2012) Setting targets for reducing carbon emissions from logistics: current practice and guiding principles. Available at www.tandfonline.com/doi/pdf/10.4155/cmt.12.62

Ministry of Transport (2016). Transportation Energy Conservation and Environmental Protection, '13th FYP' Development Plan.

Partnership on Sustainable, Low Carbon Transport (SLoCaT) (2015). Intended nationally-determined contributions offer opportunities for ambitious action on transport and climate. Available at www.slocat.org

Research and Markets (2016). Global Parcel Delivery Market Insight Report 2015. Available at www.researchandmarkets.com/research/4x499l/global_parcel

SLoCat (2015) Intended nationally-determined contributions offer opportunities for ambitious action on transport and climate change. Available at www.slocat.org

Smart Freight Centre (2015) Green freight programmes worldwide. Available at www.smartfreightcentre.org/main/info/publications

Smart Freight Centre (SFC) (2016a). Analysis of potential Smart Freight Leaders linked to Green freight programmes. Available at www.smartfreightcentre.org/main/info/publications (accessed 27 February 2018).

Smart Freight Centre (2016) GLEC Framework for Logistics Emissions Methodologies. Amsterdam.

Smart Freight Centre (2017). Barriers for carriers – Insights on barriers to adopting fuel saving technologies and measures for trucks. Available at www.smartfreightcentre.org/main/info/publications

Van den Berg, R. and De Langen, P.W. (2014) Assessing the intermodal value proposition of shipping lines: Attitudes of shippers and forwarders, *Maritime Economics & Logistics*, 17(1), pp. 32–51.

6

SUSTAINABLE BANKING

The mysterious role of commercial banks in achieving a sustainable economy

Gizem Goren and Teun Wolters

Introduction

In the wake of the economic crisis of 2008, to a greater extent than before, conventional banking seems to have an eye for adopting sustainable policies and practices. The crisis caused a profound experience of economic hardship. It made a new generation of bankers call for a serving and sustainable banking sector. However, it is still not fully clear whether the banking sector or individual banks are making real progress in these areas. Although the large banks seem to pay attention to a variety of topics that relate to sustainable development, they remain ambiguous as to the real sustainability impacts of their lending services and investment criteria.

Often, bank executives do not know yet how to mainstream sustainability and integrate it into their business models. Can we expect that the banks will play a significant role in reaching a sustainable society?

To enlighten this mysterious area, this chapter focuses on how a bank can be evaluated and positioned in terms of corporate sustainability. For that purpose, it unfolds a phase model that helps recognize how a bank evolves towards fully fledged sustainability.

The following questions take central stage:

*   How can a bank's position in sustainable development be identified?
*   Which sustainability indicators are specific to the banking sector?

To address these questions, this chapter uses the Phase Model of Sustainable Development (or the Tulder model; Tulder *et al.*, 2014). This model describes various phases that companies go through in the transition towards corporate sustainability. It also describes various tipping points when shifting towards a next phase. A great part of the research underlying this chapter was a matter of adapting this model to the commercial banking sector.

According to the Tulder model, a sustainable enterprise wishes to contribute to the good of society and acts accordingly. By doing so, it achieves more than is legally required; it is a process through which a company complies with the expectations of various stakeholders (Tulder *et al.*, 2014).

The concept of sustainable development (as defined by the Brundlandt Commission (United Nations, 1987) is the outcome of a growing awareness of the global links between mounting environmental problems, socio-economic issues such as poverty and inequality and concerns about a healthy future for humanity (Hopwood *et al.*, 2005). However, since 1987, the sustainable development literature has further evolved, dealing with issues concerning (1) the relationship between the economy, society and environment and (2) the field(s) that should be given priority.

Companies, including banks, can determine their position in terms of these issues and by so doing express their own sustainability profile. The sustainability performance of commercial banks is partly a matter of how seriously they take the sustainability issues in their strategies and policies (intentions), and partly a matter of the factual evidence of what sustainability means in the their core business (facts). Benchmarking can be helpful in evaluating the banks in these areas.

Corporate social responsibility

As of the 1960s, Corporate Social Responsibility (CSR), including stakeholder theory, has developed to counter the dire consequences of unchecked corporate power, such as environmental degradation, poor consumer products and inhumane working conditions (Logsdon and Wood, 2002). The CSR concept suggested that corporations had to focus not only on their economic performance but also on their social and environmental performance. In general, social responsibility is the obligation of decision makers to consider and protect the welfare of society along with their own interests (Davis and Bloomstorm, 1975).

There has been a dramatic increase in companies seeking to engage in discussion with civil society organizations. Many companies have promoted processes of stakeholder dialogue and engagement to increase accountability, transparency and trust (Burchell and Cook, 2006). This development can be seen as a consequence of contemporary changes in the business world, involving the rise of globalization, the dominance of information technology and an increase in societal awareness of the impact of business on communities (Freeman *et al.*, 2010). The concept of CSR offers a broad perspective to business and society relations.

Over the course of time, the literature has offered various conceptualizations of CSR. Despite differences, the various conceptions of CSR have in common the aim to go beyond mere financial criteria (Freeman *et al.*, 2010). To be socially responsive, a corporation must (1) identify its stakeholders, (2) identify their stakeholders' demands and (3) establish a dialogue with them about issues and priorities (Perez and Rodriguez del Bosque, 2014).

A stakeholder can be defined as any group or individual that can either affect the organization or is subject to being affected by the organization (Freeman, 1984).

The previous three points can be translated into the following questions: (1) How does the corporation identify its stakeholders? (2) What tools and strategies does the company choose to satisfy stakeholder demands? (3) What kind of dialogue mechanisms does the corporation use?

As stakeholders play a significant role in the Tulder model, it makes sense to delve more deeply into the stakeholder theory literature.

There is a distinction between primary stakeholders, who have a direct, contractual, official relationship with the corporation, and secondary stakeholders, who are all other stakeholders.

The primary stakeholders are mostly defined as shareholders, investors, customers, suppliers, other business partners, employees, managers and local communities, whereas the secondary stakeholders are governments and regulators (if not having any role as capital provider or governor), civic institutions, social pressure groups, media and academic commentators, trade bodies and competitors (Carroll and Buchholtz, 2008).

Stakeholder theory revolves around three main issues: (1) value creation and trade in a rapidly changing global business context; (2) the ethics of capitalism: the connections between capitalism and ethics; and (3) the managerial mindset: how to create value and explicitly connect business and ethics (Freeman *et al.*, 2010).

A key distinction is between descriptive, normative and instrumental perspectives (Donaldson and Preston, 1995). Descriptive stakeholder theory is about 'how corporations actually manage their stakeholder relationships' (Frynas and Yamahaki, 2016). Normative stakeholder theory focuses on 'how the organizations should take the legitimate interests of all stakeholders into account'. Instrumental stakeholder theory identifies the connections, or a lack of them, between stakeholder management and the achievement of corporate objectives (Donaldson and Preston, 1995).

Corporations need to identify their stakeholders and classify their importance. Stakeholders can support or harm corporations in different ways. We can distinguish between (1) the stakeholder's power to influence the firm, (2) the legitimacy of the stakeholder's relationship with the firm and (3) the urgency of the stakeholder's claim on the firm (Mitchell *et al.*, 1997); these are fundamental factors that corporations should take into consideration when prioritizing stakeholders. From this perspective, power is related to a stakeholder's ability to impose his/her will on others despite resistance; legitimacy relates to the legal and moral mandate to exert power with regard to a claim; and urgency is the degree to which a stakeholder feels an urge to call for immediate attention (Mitchell *et al.*, 1997). The most important stakeholders are, after all, the groups which unite all three factors, involving both primary and secondary stakeholders (Tulder *et al.*, 2014).

Stakeholder theory offers a new way to consider organizational responsibilities, as it suggests that corporations are not solely driven by short-term profits. Stakeholder relationships are highly relevant when examining corporate sustainability. In general, there are two main reasons for this. First, sustainable development asks for the integration of economic, social and environmental concerns (Steurer *et al.*, 2005). This kind of integration requires a great deal of collaboration between different parties in society. Second, stakeholders may be indispensable when it comes to sustainable enterprise. Therefore, companies intending to engage in sustainable business must be prepared to involve their stakeholders (Gilbert and Rasche, 2008; Tulder *et al.*, 2014).

The Triple Bottom Line approach

Recognition of the environmental and social aspects of production and doing business has resulted in the concept of the Triple Bottom Line (TBL) (also indicated as the three P's of 'People, Planet and Profit') (McWilliams *et al.*, 2014). It provides an (accounting) framework for measuring corporate performance from three perspectives: economic, social and environmental (Goel, 2010). For the most part, the environmental aspect is the impact of a business on environmental resources and what that means for future generations (Elkington, 1998).

However, social and environmental performance measures are hard to define and quantify (Slaper and Hall, 2011). The notion of (environmental) sustainability may have different meanings to different industries (Schulz and Flanigan, 2016). In the literature and in business there are studies aiming at producing indicator lists to evaluate a corporation's environmental and social performance. Sustainability indexes, such as the Ecological Footprint (EF), the Environmental Sustainability Index (ESI), the Well-Being Index (WI), the Living Planet Index (LPI, the Dow Jones Sustainability Index, the Global Reporting Initiative (GRI) and Sustainability Balanced Scorecards, are the contemporary frameworks by which one can examine and evaluate a corporation's environmental and social impacts. Even though these frameworks are criticized by some scholars and professionals for being inadequate, at present they seem to be the best available tools in the field.

When it comes to the quantification and monetization of the TBL indicators, the banking sector could play a significant role. As a vital intermediary between the world of finance and other economic activities, the banks are one of the most prominent stakeholders of the corporations. They can bring pressure to bear on their clients, if not demand from them, to make use of certain sustainability criteria, certifications and sustainability standards. In the next chapter, a bank's achievements in this field will be one of the indicators measuring a bank's sustainability level.

Sustainable development and the banking industry

The financial sector is a major player in achieving sustainable development. Financial capital is an indispensable ingredient in achieving sustainable development

(Weber and Feltmate, 2016). However, it is only since 1996 that environmental, sustainable or socially responsible investment products have begun to penetrate the market (Weber, 2012). The late entry of the financial sector's role in sustainable development was caused by a misperception of their impact on the environment (Jeucken and Bouma, 1999).

The Kyoto Protocol (1997) was a significant step in providing a foundation for the development of the first global carbon markets (Chichilnisky and Heal, 2000). Reducing carbon emissions in the signatory countries demanded new investments and financial products. Hence, the banks and the other financial institutions could not refuse taking part in this new development and provided funding for the investments. Concurrently, the United Nations Environment Programme Finance Initiative (UNEP FI) was established aiming to integrate the environmental and social dimension into financial performance and risk. According to UNEP FI, the banks should consider the impact of their various operational activities and their products and services in light of the needs of the current as well as future generations (UNEP FI, 2007). The launch of the Equator Principles (EPs) in June 2003 was the following significant step taken by the commercial banks towards a common set of environmental and social standards. The EPs provide a framework for banks to review, evaluate and mitigate or prevent negative environmental and social impacts and risks associated with the projects they finance (Polonskaya and Babenko, 2012). However, it was only in the wake of the 2008 economic crisis (caused by the financial sector) that policy makers and representatives of financial and other sectors agreed to a more balanced way of banking by taking sustainability measures and concerns into account. Therefore, banking practices had to be changed. While for long the banking sector's success was assessed almost exclusively in terms of its immediate performance, the crisis revealed that many aspects of the current, increasingly global, banking model were not sustainable (Aziakpono et al., 2014). In addition, short-termism and excessive leverage remain significant drivers of instability and reasons for ignoring longer-term sustainability-related risks in financial decision-making (UNEP, 2015). A financial crisis can accelerate new ways of doing business, taking into account shifting realities that define the market place (Lubin and Esty, 2010). The banks' usual approach towards sustainability could be called the 'business case for sustainable development' involving a limited number of sustainability-led financial products, which, however, go together with a majority of still conventional (unsustainable) financial products. Then, the focus is mainly on adopting certain niches of sustainability to manage certain financial risks and opportunities (Weber and Feltmate, 2016). The time has come for the banking sector to revise its culture, strategies, services and products and integrate sustainability in all of its products. Rather than 'the business case for sustainable development', this approach could be called 'the sustainability case for business' (Weber and Feltmate, 2016). The latter case entails the creation of not just short-term financial and economic value, but also long-term environmental and social value for a wide range of products and stakeholders (Polonskaya and Babenko, 2012). The deep-seated banking practices, motivated by familiar one-sided business models, need to be abandoned or at least reshaped to incorporate the long-term interests of society and the ecological environment.

The financial sector plays a major role in the real economy; it has a significant impact on the economy and society at large. It mobilizes savings and allocates credits across space and time. It provides not only payment services but also, and more importantly, products that enable firms and households to cope with economic uncertainties by hedging, pooling, sharing and pricing risks (Herring and Santomero, 1996). Moreover, banks transform money in terms of duration, scale, spatial location and risk and, therefore, have an undeniable impact on the economic development of nations (Jeucken and Bouma, 1999).

To fully understand the banking sector's role in sustainable development, it is important to make a distinction between the banks' impacts because of their (1) internal processes (direct impacts) and (2) products provided to clients and the conditions that clients have to fulfil (indirect impacts).

A bank's internal processes refer to the daily office operations such as the usage of paper, waste management, energy efficiency and working conditions, whereas the external processes and products refer to what a bank delivers to its clients. The environmental burden of a bank's internal energy, water and paper use may be substantial, but is overall lower than in manufacturing (Jeucken and Bouma, 1999), although many office buildings that banks make use of are still far from being sustainable.

In terms of sustainability, the external processes and banking products have a much greater impact. These can be analysed from the following perspectives:

1   The lending decision, financial flows of the banking business.
2   Sustainable products and services (innovation).
3   Client relationships and social conduct.
4   Others.

The lending decision, financial flows of the banking business

Banks have the power to decide which businesses or projects they lend money to or invest in (Weber and Feltmate, 2016; Scholtens, 2006). That fundamental right of the banks can be examined from several sustainability perspectives.

First, banks can contribute to sustainability by not lending money to unsustainable, harmful projects. Given the size, location and economic characteristics of many projects, many conventional projects tend to have significant adverse environmental and social impacts (Wright, 2012). Therefore, having in place sustainability-led lending policies, guidelines and procedures will prevent investing in businesses, sectors and projects that are not compliant with certain sustainability criteria; this, in principle, is an effective method to promote sustainable development. For example, ABN AMRO Bank N.V, active in retail, private and corporate banking, follows lending and risk management standards and principles in accordance with the United Nations Environment Programme Finance Initiative (UNEP-FI, 2007; ABN AMRO Bank N.V, 2016).

Second, financing sustainable development requires capital flows to be redirected towards critical priorities, away from polluting and unsustainable, natural resource-intensive activities (UNEP, 2015). In other words, financial capital appears to be one of the most important ingredients needed to achieve global sustainable development (Weber and Feltmate, 2016). According to the UN, new additional investments amounting to 5–7 trillion US dollars per annum are needed to reach the UN's 2020 Sustainable Development Goals (UNEP Finance Initiative, 2015). The banks, by funding sustainable energy systems, infrastructure and information technology, as well as social projects, can contribute positively and proactively to sustainable development.

Third, the activities and business models of the current borrowers can be reviewed to ensure that they contribute to sustainable development. Within this context, banks can advise their clients and direct them to implement business models that support sustainability while declining credits that would support unsustainable businesses. If this is done, clients will discover that their bank evaluates their plans not only on their financial solidity but equally on their social and environmental impacts.

To ensure sustainable lending processes, it is necessary to integrate environmental and social (risk) criteria on an equal footing with other criteria into the existing assessment procedures. Then, these procedures will ensure serious consideration of sustainability factors, such as natural resources, recycling, environmental impacts, employment opportunities, working conditions and health and safety. In this way, banks will be key actors in phasing out unsustainable business.

Sustainable products and services

Financial intermediaries, such as banks, are not just agents who screen and monitor on behalf of savers. They themselves are active counterparts offering a specific product that cannot be offered by individual investors to savers (Scholtens and van Wensveen, 2003). Banks can develop products and services that foster renewable energy, education, healthcare, sustainable water management, sustainable infrastructure and housing and the like (Weber and Feltmate, 2016). In the early 1990s, many banks acknowledged that sustainability generated business opportunities, and they started to create sustainable products in the fields of asset and credit management (Weber, 2005; Scholtens, 2007), even though these products were exceptions rather than the rule.

One of the first examples of a sustainable banking product was carbon trading. After the launch of the Kyoto Protocol in 1997, the financial sector started to develop products and services related to carbon emissions. The Prototype Carbon Fund (PCF), which was established by the World Bank in 1999, is one of the best-known financial products relating to mitigating carbon emissions, with 220 million US dollar (Weber, 2013).

In addition to carbon trading, the banks have introduced green bonds. Green bonds may involve social impact bonds, micro-finance bonds, renewable energy bonds, etc. Green funds ensure that investors' money will be invested only in green projects, such as organic farming, renewable energy, green infrastructure investments, green entrepreneurial projects or social projects. Currently, the green bond market is worth more than 100 billion euros (ABN AMRO Bank, 2016).

On the debtor side, green loans are common practice. Those types of loans are granted specifically for funding green investments, which often benefit from reduced interest rates. Investments related to energy efficiency, the reduction of air pollution emissions, improvements in environmental performance and cleaner production processes and technologies are eligible for funding by green loans. Thereby, the negative impacts of the businesses are lowered and sustainable practices embedded into existing business models.

On the household side, several banks in Europe and USA provide products such as environmentally friendly mortgages. Individuals, who are willing to invest in sustainable solutions, such as renovating their existing properties in an energy-efficient way, purchasing a property that has been certificated as green building, or in self-building projects, have an opportunity to apply for environmentally friendly mortgages, which offer discounts on the interest rates like the commercial green loans. Those types of banking products promote sustainable living and have a positive impact on sustainable construction practices. Furthermore, the products and services that target individuals are one of the key ways of raising a broader awareness about sustainability and sustainable living.

To sum up, by introducing sustainable products and services, the banking sector is able to assist the emergence and development of sustainable businesses, as well as changing the daily practices of individuals. Therefore, innovative approaches towards sustainable products are essential indicators in demonstrating the bank's position in sustainability. This will be analysed below.

Client relationships and social conduct

The banks have widespread information networks in the various segments of the economy. Moreover, banks often have access to exclusive information about a company. They use this information to judge the credit-worthiness of a client or a potential client (Goss and Roberts, 2011). Moreover, as financial intermediaries, banks not only provide finance but are also involved in project design and implementation (Haupt and Henrich, 2004). For example, early-stage entrepreneurial financing, community investing and project finance suggest that in regard to the provision of project funding, it is possible for a bank to help shape the ideas behind a project (Scholtens, 2006). Because of this unique position, banks have opportunities to influence their clients' business strategy and operations and induce them to execute sustainable practices by providing education, face-to-face meetings, and customized products and services.

Currently, there are banks in Europe which conduct such policies. For example, Rabobank, the largest green bank in the Dutch market with 1.9 billion euros in outstanding green loans, launched a policy called 'sustainable client photo' in 2015. According to this policy, Rabobank prepares a distinctive sustainability report for each client, which is composed of advices, providing insight into the business' strengths and weaknesses and the opportunities and threats in the market (Rabobank Group, 2016). The bank fosters this policy by arranging customer meetings and giving guidance to clients to rearrange and restructure their businesses in a more sustainable way.

The social conduct aspect of banking refers to all social and environmental activities in which a bank engages without considering financial return. It can concern philanthropy, presenting a set of voluntary activities such as donations to charitable causes, active participation in environmental conservation or sponsorship of cultural activities (Carroll, 1979). Today, most major banks publish such activities on their websites and in their sustainability reports.

It should be realized that philanthropy and charity have their limitations in terms of sustainability, because it is about what a company does with its profits once earned, while real sustainable value is about how a company earns its profits (Laugel and Laszlo, 2009). Therefore, these kinds of bank activities are not further considered when searching for sustainability indicators.

Others

Sustainability in financial services is not just a matter of being green or eco-friendly (Eccles and Serafeim, 2013); it is equally about being productive, transparent and responsible towards shareholders and other stakeholders, including employees, customers, counterparts and society itself (Ramnarian and Pillay, 2016).

Stakeholders involved in the provision of financial services can be grouped as employees, customers, investors, governmental and regulatory bodies, and rating agencies. However, there are other societal stakeholders as the financial industry has extensive connections with many players in society (Laugel and Laszlo, 2009).

Like any other enterprise, a bank has to build effective means of communication and consultation. Holding regular meetings with the stakeholders, working together, identifying market priorities and developing appropriate policies are ways to entertain mutual relations with the stakeholders.

Some banks seem to see stakeholders as a burden rather than as partners in doing business. However, stakeholders are indispensable actors in evaluating and improving sustainability. Stakeholders may have expertise, networks and data which can play a role in generating new knowledge. Collaboration with stakeholders and developing partnerships are common procedures in current banking practices; however, the transparency and consistency in these relationships should be much more pro-nounced. In sustainable banking, the role of stakeholders is indispensable, requiring a high degree of accountability.

Stakeholder pressure can influence a company's decisions regarding product design, sourcing, production or distribution (Parmigiani *et al.*, 2011). This also applies to banks; there can also be pressure from external stakeholders (e.g. environmental NGO's) persuading banks to reconsider certain lending decisions because of the indirect social and environmental impacts of their decisions. For their part, banks can put pressure on their clients to pursue or refrain from certain policies.

For example, ABN AMRO has subscribed to the principle of Free Prior and Informed Consent (FPIC). Like the need to respect all other human rights, this principle is enshrined in the bank's sustainability policy. Based on that principle, in 2017, 'ABN AMRO initiated an engagement process with ETE (a company with which it has financial relations) in which it expressed its concerns about the involvement of its subsidiary ETP in the Dakota Access Pipeline Project (in the USA), and emphasized that ETE should exert pressure on ETP and the other project sponsors to reach a solution that is acceptable to all the parties impacted by the construction of the pipeline, among which the Standing Rock Sioux Tribe' (source: www.abnamro.com; accessed 29 December 2017). As ABN AMRO was neither involved in the construction of the pipeline nor was a direct lender to the project sponsor, it was its indirect relationship with the project, through ETE, that incited the bank to take action (source: www. abnamro.com).

The Tulder model and its phases

The Tulder model and phase-based indicators

The phase model of sustainable development is used to analyse the position of a corporation in regard to sustainability. According to the model, there are four main phases that a corporation goes through in the development of sustainable policies and practices. Those phases present a combination of two key behavioural dimensions:

1   A company's basic attitude to society and societal issues.
2   A company's measure of societal responsiveness.

A company's basic attitude to society and societal issues

On the 'basic attitude' dimension of the model, a company can have two main attitudes towards sustainability. According to the first attitude, a company is inclined to consider sustainability a 'liability'. Then, legal requirements guide business behaviour; efforts mainly focus on avoiding legal claims (Tulder *et al.*, 2014). The second attitude is labelled as 'responsibility'; here, the corporate leadership accepts responsibility for the effects of its corporate strategies on society and acts accordingly.

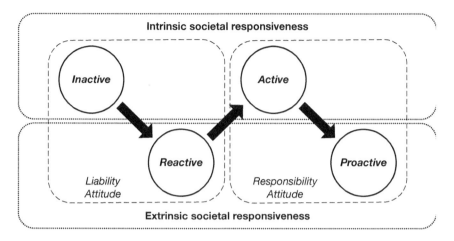

FIGURE 6.1 Overview of the Tulder model

A company's measure of societal responsiveness

On the 'societal responsiveness' dimension of the model, companies are categorized according to their motivation; there are 'intrinsically' or 'extrinsically' motivated companies. If a company follows its own ideas on sustainability without much external consultation, it is intrinsically motivated (Tulder *et al.*, 2014). On the other hand, a company can be strongly influenced by certain external groups – key stakeholders – such that this external influence shapes its corporate sustainability strategy and related practices.

According to the model, the companies go through four main phases, which are indicated as inactive, reactive, active and proactive. Figure 6.1 provides a stylized overview of the model.

The four phases of the Tulder model

Companies *in the inactive stage* believe that sustainability is a prime responsibility of governments and customers; therefore, for them, sustainability is a matter of doing what is prescribed by law (if sufficiently enforced) without having any intention to go a step further (Tulder *et al.*, 2014). These companies' main concern is generating (maximum) profits; it is a strongly inward-looking (inside-in) business perspective aimed at efficiency in the immediate market environment (Tulder and Fortainer, 2009). These kinds of companies tend to execute conventional business models.

In the reactive stage, sustainability is about a limited range of environmental and societal issues. Even though a reactive company's mission or vision statement may mention aspects of sustainability, there is no corporate strategy that addresses sustainability issues. Rather than intrinsic motivation, it is extrinsic motivation, that is, external pressure, that prevails. That means, it is the signals from customers and other stakeholders relating to the company's reputation that rouse a company's interests in certain aspects of sustainability. For instance, such signals may cause a company to introduce green(er) products and services and publish sustainability reports. However, in this stage there remains the risk that the motivation never really becomes intrinsic. Here, sustainability tends to stay dependent upon 'conditional morality' in the sense that managers only 're-act' if and when competitors do the same (Tulder and Fortainer, 2009). Here, there is also the danger of 'greenwashing'.

The active company considers sustainability a worthwhile goal because it is good and makes *sense per se* (Tulder *et al.*, 2014); it is inspired by ethical values and virtues (Tulder and Fortainer, 2009). Such a company sets specific sustainability goals, defines core themes and programmes and presents the results to its stakeholders. Business motivation is built on the question of 'where can we make a difference to society?' (Tulder *et al.*, 2014). An active company not only exchanges ideas with its stakeholders but also provides the means to prepare for and carry out joint efforts in the field of sustainability. In this phase, stakeholders are acknowledged as (potentially) beneficial partners. As sustainability is considered a corporate responsibility, an active company's product and service innovation is often based on sustainability goals. Active companies will publish well-structured and balanced sustainability reports, mostly using the guidelines of the Global Reporting Initiative (GRI). In the active phase, the supply chains will receive ample attention. Therefore, suppliers are audited and action is taken accordingly. Also, engagement and collaboration within the supply chain is part of the active company's strategy and how it is implemented.

In the proactive stage, sustainability is fully integrated into a company's strategy, while all the operators and executives are aware of this strategy. As a consequence, the proactive company's business models are wholly based on sustainability goals. Here, strategic investment decisions are primarily based on visionary leadership, not on a convincing financial underpinning (Tulder *et al.*, 2014). A proactive company differs from other company types because it gives precedence to trans-formational sustainability goals over all short-term 'business as usual' considerations. The ultimate goal of a proactive company is to provide sustainable products and/or services to its customers; therefore, it is not preposterous to close down a product line even though the product is still bringing profits to the company. This stage is the ultimate stage in which sustainability is a part of every aspect of the company with full involvement of stakeholders as supporters or partners (Tulder *et al.*, 2014). Companies in this stage are industry leaders. As sustainability is fully integrated into their business models, they tend not to publish separate sustainability reports, but cover all the data related to their sustainability performance in their integrated annual reports.

The Tulder model provides general indicators to measure the level of corporate sustainability according to each stage. The model can be developed further to assess a bank's position in sustainability (see Table 6.1).

TABLE 6.1 Indicators of corporate sustainability

Stages	Business Case	Vision on Sustainability	Business Motivation	Applications of Sustainability	Sustainability Reports	Relations with the Stakeholders	Customer and Supplier Relations
Inactive	Classic	No sustainability vision.	Profit maximization. Maintaining the status quo at any cost.	Limited to meeting legal requirements.	Published if legally required.	Primary contractual. Seen as a burden if there is no direct interest. Monologues.	Unless there is a commercial interest, customer demands are ignored.
Reactive	Defensive	Focus is on the environmental and societal issues. Very general.	Reputation, cost advantage, direct market response.	Connection with the corporate strategy is weak. External pressure is determinant.	Reports published but full and transparent.	Endeavour to earn official recognition from the stakeholders.	Reacts to customer demands and introduces standards.
Active	Strategic	Sustainability is necessary because it is "good". More is possible and necessary.	Where and how can we make a difference to society?	Physical audit of the supply chain. New product innovation.	Full reports. Global Reporting Initiative (GRI) guidelines are followed.	Open dialogue, exchange ideas, collaboration with all stakeholders.	Innovation of sustainable products or services. Cooperation in the supply chain.
Proactive	Societal	Sustainability is fully integrated into company strategy.	Societal value over financial value. Visionary leadership.	Industry leader, long-term strategy. Unsustainable businesses is shut down.	Sustainability report is an integrated part of the annual report.	Explore new horizons and collaboration at a strategic level. Equality and reciprocity between all parties.	Partnership with NGOs. Helping suppliers towards sustainability. Co-creation with all parties.

Internal and external drivers of the Tulder model

The Tulder model distinguishes three different kinds of internal driver:

1 A need for sufficient harmony between company departments to enable the alignment of sustainability policies.
2 The various perception gaps between employees and the official organization that may thwart a smooth transition from one phase to the other.
3 The leadership challenge (sustainability requires convinced senior managers whom the employees trust and feel inspired by).

Moreover, the model recognizes three main external drivers: international coordination, special issues and stakeholders.

International coordination

In regard to international coordination, companies that operate overseas can also contribute to the adaptation and enhancement of sustainability policies in their supplies chains, even in areas where local governments and business standards are less demanding. This can be stimulated by public pressure in the country where a company has its head office or otherwise has an established position.

Special issues

The special issues are of two kinds, that is, there are internal and external issues. On the internal side, the model suggests that the more a company integrates its relationship with the secondary stakeholders, like NGOs, the more the employees of the company will be open to change and support the implementation of sustainability policies. On the other hand, companies that have a defensive structure are inclined to resist change. However, there may happen sudden, unexpected events (e.g. protests, calamities) that cause a breakthrough and accelerate the transition towards (a next phase of) sustainability.

Stakeholders

The third external driver of the model is composed of a company's stakeholders. According to the Tulder model, the primary stakeholders who are directly in contact with the organization's functional departments are the most important driver (or barrier) (Tulder et al., 2014). The first group of primary stakeholders consists of the employees of the corporation. As discussed before, in general, the engagement of employees is essential to a successful path towards sustainability. The second important stakeholder group in a transition is composed of suppliers, consumers and governments.

Collaboration with consumers can be an important driver of sustainability; uniting experts and consumers in generating smart ideas for making sustainable products often increases the likelihood of being successful (BBMG, 2012).

Phases of the Sustainable Banking Model

The four phases

This section discusses the various phases of the Tulder model as applied to the banking sector.

The defensive phase

In the defensive phase, banks tend to deal defensively with legal requirements and regulations. For them, designing and implementing sustainable policies and practices is beyond their scope. Their internal and external processes as well as their banking products are conventional only. Defensive banks operate for the purpose of short-term profits and increasing their market share. Contributing to social and environmental improvements as such are not regarded as part of their business. Any project in which they invest has to offer a satisfactory, if not maximum, financial return. A bank perceives its own contribution to the common good as its capacity to generate financial returns for its shareholders (Arnsperger, 2014). In addition, defensive banks believe that a bank cannot be held responsible for its clients' actions. This standpoint is reflected in their external processes and banking products. Moreover, rather than controlling resource consumption, waste management and ecological footprint issues as a means to promote efficiency and save costs, defensive banks take these costs for granted.

The reactive phase

In the reactive phase, a bank considers sustainability as a 'liability'. However, unlike defensive banks, a reactive bank takes business trends into account, primarily motivated by cost savings, stakeholder pressures and reputation concerns. Therefore, it begins to adopt sustainability policies. For example, ecological footprint-related issues, resource consumption and waste management issues could be put on the policy agenda leading to new lending criteria or slightly modified banking products. However, the attitude towards sustainability remains pragmatic and *ad hoc* in nature under the banner of short-term profit maximization.

Here, the relationship with the stakeholders is rather unstructured or fragmented; communication with them is neither systematic nor strategic. It happens only when there are incidental pressing circumstances.

However, reactive banks aim to demonstrate their positive responses to sustainability in a visible way so as to boost their reputation. This can be done by activities such as publishing sustainability reports, advertised messages and charity projects. Reporting is a major instrument in a reactive bank's reputation management. In this phase, the sustainability reports do not demonstrate structural change; only minor achievements – either in internal operations or in a few selected greening projects (preferably supported by government subsidies) –

are presented. Such minor achievements are often presented as the height of sustainability.

The active phase

In the active phase, sustainability is perceived as a corporate responsibility rather than a liability. A responsible bank is committed to a structural transformation towards sustainability. This attitude is basically internally motivated and supported by explicit sustainability policies and procedures which include both internal processes and external banking products. Therefore, the active phase inaugurates a real turning point in sustainable banking.

In this phase, a bank's first aim is to reconceptualize its corporate values and make sure they are known throughout the organization. To accomplish that, it is crucial to involve the bank's employees. Providing training in sustainability and developing certain business skills comprise the first steps. Here, it is advisable to develop adequate training programmes together with the employees to increase motivation and commitment throughout the bank. Moreover, company newsletters and bulletin boards can be used to spread information on the bank's sustainability policies.

Stakeholders are acknowledged as a component of the process. For that reason, there are regular meetings with the representatives of different stakeholder groups, which a bank asks for advice and feedback in the area of sustainability.

In the active phase, the most important change takes place in the external processes and the banking products. A commercial bank's *raison d'être* lies in being an intermediary between the cash from savings and investments. Therefore, the most powerful impact that a bank can have on sustainability lies in the criteria it uses when deciding on giving loans or investing in certain projects. Sustainability should be embedded in these criteria.

Moreover, an active bank can innovate and present new sustainable banking products and services for both savers and investors. This attitude is also a matter of prioritizing sustainable businesses and investments seeking funding. The ultimate purpose of an active bank is providing means and tools to the actors of the economy who wish to contribute to sustainability. In other words, it is about converting business as usual into sustainable business. Therefore, a bank's approach to product innovation, risk assessment and funding is of vital importance.

A bank in the active phase begins to support sustainable product and service innovation in two regards. First, it will increase the resource efficiency of its own processes. Second, an increasing part of an active bank's business will derive from better products and services. Green loans, bonds and mortgages are just a few examples of sustainable banking products. Environmental and ecological investment plans will gain access to financial capital more easily if banks distinguish and appreciate sustainable banking products.

In this phase, channelling capital into less harmful businesses is the main objective of the modified risk assessment process. As a consequence, ethical, societal and environmental concerns will penetrate lending policies and procedures; manuals are rewritten accordingly. Sustainability becomes part of the rating and scorecard systems. Having a credit risk assessment model that involves sustainability is a strong indication that a bank takes sustainability seriously.

Projects in areas such as clean energy, resource efficiency, environment protection, green transport and sustainable agriculture are likely to fit with the sustainable agenda of the active bank, presenting new market opportunities. New social and environmental guidelines will be published. However, in this phase it is unlikely to gain a full insight into an active bank's credit portfolio because of a lack of transparency. Because of that, it is not possible to fully assess how much of an active bank's entire business deserves to be called sustainable. Furthermore, for outsiders, there remain unknown areas as to a bank's risk assessment methods, the weight of the environmental and social risks, how calculations are done under the rating system and on what grounds sectoral differentiation is carried out. Moreover, active banks do not normally disclose their plans for their future portfolio management. Therefore, it is hard, if not impossible, to predict when unsustainable lending will be phased out.

As a consequence, even though the active phase can be recognized as the first substantial step towards sustainable banking, there is more to achieve. For that matter, an active company must develop into a proactive organization in making sustainability normal practice.

The sustainable/proactive phase

The sustainable/proactive phase indicates the ultimate approach within the current economic system. As long as sustainability is a far cry from its reality, sustainable/ proactive banks are pioneers and holders of essential transformative capabilities.

The main concern of the proactive bank is leading in the transformation of conventional ways of doing business towards sustainability. Here, responsibility as a basic attitude is supported by well-designed business plans, covering both a bank's internal affairs and its external policies and practices. A sustainable bank's value proposition must integrate ecological, social and economic value through its products and services (Schaltegger et al., 2015).

The sustainable/proactive phase begins with significant behavioural change among a company's top and middle management. It is a state of maturity grown from the active stage. This shift in mentality is fed by increasing demand and expectations emerging from various stakeholders, such as employees, governments and NGOs. It is more and more understood that previous attempts to bring sustainability closer to realization may have had their merits but are insufficient to reach the ultimate goal. Therefore, the time has come to fully endorse the transformation needed and take action. Banks that take part in this new awareness and act accordingly can be identified as belonging to the sustainable/proactive phase.

A bank entering the sustainable/proactive phase will revise almost all key facets of its business. Such a bank has to implement significant structural changes; its employees have to accept and support these changes. The senior management of such a bank must show leadership in carrying out the necessary cultural and structural change to ensure it is supported by all segments of the organization.

Sustainable/proactive policies regarding internal processes

Three different kinds of proactive policy

The following three proactive corporate policies can be identified: (1) specific targets; (2) HR policies that support the company's sustainability agenda; and (3) engaging in sustainability assessment, monitoring and reporting.

Specific targets

A sustainable/proactive bank defines specific targets for its own internal processes to minimize its negative impacts on the environment (think of waste, material use, energy, water consumption, reduction of CO_2 and responsible office furniture and equipment). Moreover, a proactive bank uses inclusive calculation methods and annual targets for each area for which environmental and social improvements are envisaged.

HR policies that support the company's sustainability agenda

A proactive bank shapes its recruitment policies in a way that conforms to its proactive agenda. Fully enabled and motivated employees demand progressive HR policies that combine high performance requirements with fairness, trust and transparent policies (Wolters, 2013, chapter 5). These may involve: decentralized decision making (in teams), diversity, measures to ensure a satisfactory work-life balance, fair individual performance measurements and salary policies, and appropriate training programmes.

Engaging in sustainability assessment, monitoring and reporting.

This is to be restructured on the basis of transparency; disclosure of the bank's short- and long-term objectives and their consistency is crucial. The annual reports should include all the information related to both achievements and setbacks. Proactive reporting is about the following content:

1   Information about the bank's risk assessment policies and lending decision criteria.

2   Declaration of a bank's explicit environmental policy and a summary of its sustainable management performance.
3   The analytical assumptions relative to the rating systems.
4   Detailed portfolio analysis based on sectoral and regional information.
5   Verifiable data about the bank's outstanding achievements and a breakdown based on sustainable versus unsustainable business.
6   Annual and long-term targets regarding the phasing out of unsustainable businesses and an explanation of the rationale behind these targets.

About external products

The most distinctive characteristic of the sustainable/proactive bank lies in the sustainable products and services it provides to its clients. The indirect impacts of banking products are concerned with the most important social and environmental issues with which the banking sector is expected to be concerned (Vigano and Nicolai, 2009).

Innovation

In a market system, sustainable development requires sustainable innovation, especially to involve mainstream customers (Schaltegger and Wagner, 2011). For a sustainable/proactive bank, innovation is not an isolated activity but involves a variety of departments, employees and external stakeholders. It is conceivable that proactive banks are involved in discussions on how sustainability-led innovation can be shaped in a way that involves large segments of society. For instance, proactive/sustainable banks can intensify discussions on the circular economy and sustainable food and what they could mean to sustainable business.

Advisory service/transformative influence

A proactive bank is aware of the fact that under the present economic circumstances not all conventional business can be abandoned overnight. For that reason, it sets mid- and long-term portfolio objectives and gradually lowers its role in conventional business.

In this context, a sustainable/proactive bank can be expected to offer advisory services. For instance, banks can help explore new markets and develop elements of a sustainable business plans for its clients. As banks often have a comparative advantage in knowledge (regarding sector-specific information, legislation and market developments), they can fulfil an important role in reducing the information asymmetry between market parties (Jeucken and Bouma, 1999). Moreover, a proactive bank can persuade its clients to give sustainability a high priority and can follow that up by informed advice.

The proactive bank prepares tailor-made periodic reports for each client. These hold general information regarding the sector, sustainable business opportunities in the sector, future trends and market forecasts. The reports also recognize the unsustainable aspects of the current business along with the feasibility studies for the required investments and funding options. In this way, the clients are stimulated to embrace a sustainable business approach and carry out the necessary investments.

In addition to reporting and face-to-face meetings, the remodelled client relationship involves promoting mutual contacts between a bank's clients. As sustainability requires innovation at different loci in the product lifecycle, good contacts between clients operating in different industries has the potential of creating new sustainable business opportunities. As a proactive bank possesses a great deal of information about industries and markets, it can advise clients and introduce them to other clients. In such a way, a proactive bank facilitates its clients in exploiting sustainable commercial opportunities. This approach is effective, especially, when it brings together a start-up or low-capital enterprise with a sustainable investment idea and other clients interested in creating sustainable solutions. Relevant areas here are business models based on the principles of the circular economy and sustainable energy (Walter, 2016).

Risk assessment and lending policies

Lending is one of the original tasks of banks. In that context, accurate identification of the environmental and social risks is important. For that purpose, a bank's overall risk assessment policies and its lending policies need to be aligned. Therefore, banks systematically screen the environmental and social performance of their borrowers. This approach involves (1) the determination of non-financeable sectors, (2) sector-specific due diligence guidance and (3) adjustment of rating and scoring systems involving sustainability. This integrated approach of lending is part of a sustainable/proactive bank's core business decisions and management.

The sustainable lending approach entails the following:

1   The relationship with the clients operating in the non-financeable sectors is frozen.
2   Clients receive sector specific guidance and policies, highlighting the unsustainable aspects of the business/investment. Following that, a sustainability framework, with specific targets, is set and discussed with the client, including a specific time frame. Within this time frame, the bank sets emission and exposure limits and manages its credit structure on that basis.
3   The proactive bank monitors its clients' improvements and adjusts its acceptable upper limits and other criteria accordingly. By so doing, a bank's portfolio is gradually cleared from unsustainable business risks.

For that purpose, a bank's senior management sets portfolio targets. These targets and deviations from them are published in the annual report (see Figure 6.2).

The organizational structure of the lending and asset management is as follows (Figure 6.3):

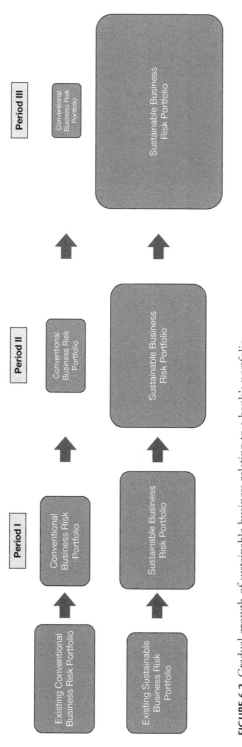

FIGURE 6.2 Gradual growth of sustainable business relating to a bank's portfolio

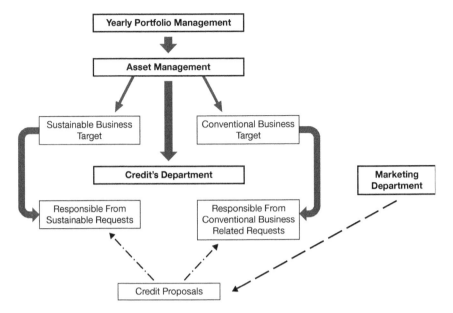

FIGURE 6.3 The organizational lending and asset management structure

Furthermore, it can be expected that such a dedicated bank will find ways of supporting sustainability-led innovative start-ups and upscaling innovative business, including micro-finance for societal projects. Consequently, the bank will be a 'change agent' for the entire economy.

Screening a borrower's environmental and social performance, including its supply chain, is one of the key features of the sustainable/proactive bank. For a proactive/sustainable bank, a sustainability risk assessment of a client's suppliers may be a tough task, especially when a client is engaged in transnational operations. For that reason, certification of goods and services by independent institutions can be a great help, such as for the following standards: ISO 14001, ISO 9000, EMAS (Eco-Management and Audit Scheme), CESCO (Certified Environmental and Safety Compliance Officer), Carbon Trust Standard, Fair Trade Certification, Utz Certified, Rainforest Alliance Certified are prominent examples among many other forms of certification that are widely accepted. The proactive/sustainable bank requires its clients to make appropriate use of these certification possibilities to harness their sustainable business.

Case Study: Rabobank and Food and Agribusiness Sector

Case study in four steps

This section presents a case study on Rabobank, in which the developed phase model of sustainable banking has been applied and evaluated. Rabobank is a Dutch (but internationally operating) commercial bank. The case study was conducted in four steps:

1 Design the case study.
2 Conduct the case study.
3 Analyse the materials that emerge from the case study.
4 Write down the findings, draw conclusions, make recommendations and indicate their implications (Yin, 1994).

In accordance with these four steps, the case study is designed with the intention to evaluate and test the previously developed sustainability model for the banking sector. Rabobank operates in 43 countries involving various banking services such as retail, whole banking and real estate services. The bank has had sustainability on its agenda since the early 1990s.

To make the case manageable, one sector was selected, namely, food and agribusiness. The bank's historic expertise and experience in food and agribusiness supports this choice.

To shape the case study, firstly, the bank's overall policies and practices were identified based on its publications and other materials. Two interviews played a central role in this study.

The first interview was with a manager of an independent Dutch initiative called De Eerlijke Geldwijzer (Fair Finance Guide), which aims to give online insight into the investment policies and practices of 10 Dutch banks and 10 Dutch insurance groups, involving 19 issues, such as climate change and human rights (EerlijkeGeldwijzer, 2015). The second interview was held with a food sector specialist/manager of Rabobank Salland.

Besides what emerged from the interviews, official documents and messages of the Fair Finance Guide and Rabobank were taken into account. The bank's sustainability performance was examined and evaluated by means of the previously developed sustainability model for the banking sector. Practical recommendations related to the bank's sustainability policies and practices were given. Finally, the outcomes and validity of the application of the sustainability model were evaluated.

De Monitor Fair Finance Guide (De Eerlijke Geldwijzer)

The Fair Finance Guide began in 2009 as a collaborative project of Oxfam Novib, Amnesty International, the Dutch labour union federation FNV, Friends of the Earth Netherlands, the Dutch Society for the Protection of Animals and the peace organization Pax.

De Monitor assesses the investment and financial services of banks with respect to 9 main international sustainability themes: climate change, human and labour rights, arms, health, nature, animal welfare, taxes and corruption, transparency and accountability, and bonuses (EerlijkeBankwijzer, 2014). An interview with a representative of the Fair Finance Guide was conducted for two reasons; first, to gain a first impression of how Rabobank's policy and practices relate to the principles of sustainable banking; second, to acquire statistical information on how the Dutch banks compare in terms of sustainability and to understand the underlying justifications. Through the interview, the prevailing issues of sustainable banking and the position of the case bank, Rabobank, compared to other Dutch banks, were identified.

De Monitor's scores give an insight into Rabobank's sustainability position in the Dutch banking sector. To conduct a reasonable and fair analysis, Rabobank's performance is compared to ABN AMRO and ING Bank as their size, business operations and spheres of influence are similar.

According to Table 6.2 (on a scale from 1 to 10), ABM AMRO gained the highest average score for sustainability performance as compared to both Rabobank and ING Bank; Rabobank takes second position. According to the ratings, Rabobank's sustainability policies on climate change, the financial sector and housing and real estate are less auspicious. In the area of housing and real estate, De Monitor suggests that Rabobank should require its clients to use certified construction materials. In the financial sector, Rabobank can do business only with financial institutions that comply with the UN Global Compact standards. In regard to climate change, the bank is criticized for not submitting its credit/

TABLE 6.2 Sustainability performance of Rabobank, ABN AMRO and ING Bank (2016)

Issues	Rabobank	ABN Amro	ING Bank
Climate Change	3	3	2
Labor rights	9	8	9
Weapons	7	9	6
Taxes	5	4	4
Animal welfare	6	7	5
Corruption	8	7	8
Human Rights	8	9	7
Bonuses	5	5	4
Transparency	5	5	4
Fishing	6	4	6
Forestry	5	6	5
Financial sector	3	3	3
Electricity production	5	6	5
Nature	6	8	6
Health	6	8	6
Nutrition	6	8	6
Mining	6	6	6
Oil and gas	6	7	5
Making industry	6	7	6
Gender equality	4	1	1
Housing & Real Estate	3	5	1

http://eerlijkegeldwijzer.nl/bankwijzer/beleidsscores/

client portfolio for measurable CO_2 reduction requirements. Moreover, it continues to finance coal mining, gas, coal and oil power plant projects as well as companies which have ties with these environmentally harmful sectors. However, De Monitor also noted Rabobank's financial support to sustainable energy generation investments.

Furthermore, Rabobank scored highest in the areas of labour rights, fighting corruption and human rights; here, it executes socially responsible policies based on international standards. Nonetheless, the bank is criticized for not conducting sustainable policies towards third parties in supply chains, involving subcontractors, landowners, etc.

As a result, despite investing in sustainable businesses and establishing sustainable banking policies, the bank's performance is mostly rated as 'sufficient (6)' or 'doubtful (5)'. The interviewee of De Monitor also highlighted the importance of this subject. According to him, fundamental violations of human rights, infringements of social justice and ecological damage occur mostly in the companies' international supply chains. If the bank were to do better it should act proactively, demanding that its clients act responsibly in their transnational operations as well. Here, the use of certified materials and certification of the suppliers' operations are offered as concrete solutions. This perspective presents a result-oriented, objective, uncontroversial formula aimed at preventing unsustainable practices in transnational businesses. For that reason, this indicator (certification) has been added to the monitor as one of the aspects of being a proactive bank.

Sustainability approach of Rabobank

Rabobank was founded in 1898 with the conviction that the cooperative concept of farmers working together would be of benefit to all (Rabobank, 2017a). The bank has a cooperative structure, and thus has no shareholders only members. The bank concluded 2016 with a net profit of €2 billion, €424 billion of issued loans and €348 billion of savings held by its customers (Rabobank, 2016). The bank claims to be the leading food and agribusiness bank worldwide, involving 85% of Dutch farmers and 40% of the Dutch food and agribusiness value chain. Furthermore, the bank's global food and agribusiness loan portfolio is worth €92.3 billion (Rabobank, 2015).

Prior to highlighting specific sectoral policies, this section examines the bank's overall sustainable development approach, including the indicators according to the sustainability model for the banking sector as presented in the previous chapter.

Based on Rabobank's 2016 Annual Report (Rabobank, 2017a) and other reports published on the bank's website, Table 6.3 summarizes its strategies and policies.

The bank's official publications on its sustainability policies and applications can be summarized as follows:

TABLE 6.3 Rabobanks's strategies and policies

Subject	Policy/figures	Explanation/integration with sustainable development
Sustainable Development Goal (The Vision)	Offering financial support, sharing knowledge and connecting parties.	Accepting co-responsibility for achieving the 17 sustainable development goals set by the U.N.
The Mission	Substantial contribution to welfare and prosperity in The Netherlands and to feeding the world sustainably.	
Stakeholder Approach	Creating value for the stakeholders and society at large.	Sustainable value is created through enduring relationships with all stakeholders and responding to customer needs efficiently.
Stakeholder Engagement	Customer feedback platforms, customer and employee surveys, dialogue with social welfare organizations.	Various communication methods are applied covering different subjects. Two-way dialogue with employees, business partners etc.
Dow Jones Sustainability Index Rating	2016 2015 Ranking: 7 5 Overall Score: 91 87	No clear explanation about the decline.
Reporting	Integrated annual report, covering sustainable development issues (GRI standards).	Invested in increasing data quality, higher reporting standards and transparency levels of reporting.
Total sustainable finance figures (Credit Portfolio)	2016 2015 2014 18.8★ 19.2 19.5 ★billion Euros	No clear explanation about the decline.
Weight of the sustainable portfolio over total credits	2016 2015 2014 4.43% 4.52% 5.09%	No forecast/information about the future target. No explanation about the decline in 2016.

Sustainable Banking Products	Green bonds, green savings, green deposits, socially responsible deposits	
Risk Strategy and Goals	Protect profit and profit growth, maintain a solid balance sheet, protect identity and reputation, healthy risk-return decisions.	Mainly conventional (financial) risks are considered as part of the Rabobank's risk strategy and goals.
Risk Culture	Careful consideration of risk/return trade-offs and appropriate measures based on up-to-date risk analysis.	Additional principles providing guidance on risk culture: integrity, respect, professionalism, sustainability, long-term relationship with customers.
Definition of Risk Appetite	Return on capital, credit risk, market risk, liquidity risk etc.	In 2016, sustainability was integrated into the risk proneness for the first time (no further detail).
Credit Risk Management/Models	Internal models to assess credit risk: probability of default, loss given default, exposure at default parameters.	Credit approval and measuring risk exposures are based on these measurement models.
Credit Acceptance Policy	Careful assessment of the customers and their ability to repay the loan that was issued.	The credit policy is developed aiming to minimize the repayment risk. However, without any detail, it is stated that the sustainability guidelines have been established for use in the credit process.
Rating Models	Sector, country, counterparty and financial risks are evaluated.	Track records of customer behaviour, financial and market data are mentioned.
Transparency of the Rating Models	Bank statement: Assumptions used in our model are not disclosed as these are considered proprietary.	The weight and importance of the variables are hidden. It is problematic to analyze is the lending approach of the bank.

2016 Annual Report (Rabobank, 2017a)

1   The bank publishes single, integrated annual reports, sustainable development issues and its sustainability policies are grouped together in one section.
2   There is limited detailed information about the bank's sustainable development policies and future sustainability agenda in the sections about its core businesses, such as lending and risk assessment.
3   Risk is defined in conventional financial terms and ratios: return on capital, liquidity risk, market risk, default risk, etc. Evidently the rating models are developed according to these variables. In the report, the effort of integrating sustainability and corporate governance factors is seldom mentioned. It is obvious that no systematic and elaborated sustainability risk assessment model has been developed to date.
4   Even though the sustainable credit portfolio is €18.8 billion, when this is compared to overall loan portfolio the sustainable lending constitutes only the 4.43%. There has been a slight decline in sustainable loans over the last three years, but no explanation is given for this decline. Also, there is no formulation of a numerical or proportional sustainable portfolio object.
5   The variables behind 'sustainable business and/or investment' are not (or only partially) disclosed. The bank's website provides a list of sustainable investment projects, without revealing what kind of indicators and measurements are used to define the 'sustainable credit portfolio'.
6   The importance of engaging with the stakeholders is highlighted in the reports. Various communication methods are used to increase dialogue. Furthermore, the roundtables in which the bank participated are published in detail.
7   The bank's sustainability performance was voluntarily evaluated; here the Fair Finance Guide was one of the initiatives. According to its own evaluation, Rabobank's sustainability performance significantly improved in 2015 but this development was not maintained in 2016. Arms, climate change, forestry, food and fisheries are among the sectors that have contributed to the final outcome. Animal welfare and remuneration are the only two fields where the bank improved its 2015 sustainability score.

Table 6.4 shows a detailed evaluation of Rabobank's level of sustainability; it is also necessary to analyse its sustainability policy framework and consider other relevant documents.

Sustainability policy framework of Rabobank

Rabobank describes sustainable development as the quality of not being harmful to people, communities, environment or natural resources, and thereby supporting a long-term social and ecological balance (Rabobank, 2016). Moreover, the bank indicates its responsibility for and commitment to lowering its direct and indirect impacts as a financial institution. Here, the term 'sustainability risk' is used as a term covering the bank's operations (potential negative impacts) and the

TABLE 6.4 Evaluation of Rabobank's level of sustainability, 2014–16

Fair Finance Guide

	2016	*2015*	*2014*
Arms	7	8	8
Human Rights	8	8	8
Climate change	3	5	3
Labour rights	9	9	9
Remuneration	5	4	2
Animal welfare	6	4	4
Health	6	6	1
Taxes and corruption (until 2015)		5	2
Taxes	5		
Corruption	8		
Nature	6	7	3
Manufacturing	6	6	1
Mining	6	6	4
Power generation	5	5	1
Forestry	5	6	4
Food	6	7	5
Oil and gas			
Fisheries	6	7	6
Financial Sector	3	3	1
Transparency and accountability	5	6	5
Housing & Real Estate	3		
Gender Equality	4		

Rabobank, 2017a.

performance of its clients and the consequences of their actions. Moreover, Rabobank supports its clients in achieving a sustainable development by providing financial solutions, insights, knowledge and access to networks (Rabobank, 2017b). In the light of these statements, the bank is aware of its role as an intermediary; it realizes its transformative influence on its clients. Here, the bank refers to its involvement in various national and international roundtables and seminars and its efforts to increase awareness among both its clients and society at large. It attempts to bring together business owners and other stakeholders, such as universities, to establish networks, including supply chains, that foster sustainable development.

To carry out these objectives, the bank follows a sustainability policy framework, which has a top-to-bottom structure:

Furthermore, the bank has decided not to finance a number of business sectors as these do not fit with its business strategy and core values. Rabobank has also expressed its commitment to sustainability initiatives such as the UNEP Finance Initiative and UN Global Impact (2016).

It can be concluded that Rabobank has acknowledged the necessity of applying a holistic banking approach to enhance its sustainability level and that of its clients.

Sustainability Policy: General framework of commitments and expectations.

Core Policies: Environment, human rights, labour standards, anti-corruption.

Theme Policies: General framework of the policies regarding animal welfare, biodiversity, gene technology, investing in agricultural commodity derivatives, land governance.

Sector Policies: armaments industry, biofuels, cocoa, coffee and cotton, extractive industries, fishery, forestry, livestock farming, palm oil, ship recycling, soy, sugarcane are defined as socially and environmentally sensitive industries and supply chains (Rabobank, 2016).

FIGURE 6.4 Rabobank's sustainability framework

Rabobank, 2016

However, after examining the bank's sustainability framework and other related documents, some fundamental questions are in need of a clear answer:

- What is its definition of a sustainable business/investment?
- What is the weight of sustainability risk in its credit evaluation method?
- What kind sustainability indicators are integrated into its rating models? And how?
- What will be the weight of a sustainable credit portfolio in the future? Is there any specific intention/policy?
- What kind of actions are taken to control and monitor unsustainable actions of the clients?
- Which product or service certifications are its clients and their supply chain partners required to possess?
- What is its strategy for creating sustainable banking products and services?
- What is its innovation strategy?
- What are successful examples of the bank's transformative influence on its clients?

Considering these factors, according to the phase model for sustainable banking, Rabobank can be ranked somewhere between the reactive and active phases. According to some aspects, such as stakeholder relationship/engagement, an overall responsible approach towards sustainable development, availability of sustainable banking products, sustainable development frameworks defined by universal values, transparency efforts and reporting standards the bank is categorized in the active phase. On the other hand, especially due to a mostly conventional strategy and risk orientation, the bank can be classified as a reactive bank; this judgement is based on the following: (1) protection of the bank's short-term profitability; (2) a risk culture based on risk/return trade-offs; (3) a risk proneness determined by financial ratios such as return on capital, liquidity risk, market risk etc.; and (4) lending decisions based on financial return, default and market risk assessments, balance sheet strength, etc. It seems that the bank is in a transition period that heralds the integration of sustainability into its core banking activities. The bank's reports are clear on its sustainability agenda, its role and policies. However, actual examples supporting the implementation of these reports and their strong statements are, as yet, not convincing. Apparently, the bank's deep-rooted corporate culture (>100 years), long-term relationships with its clients as well as viability concerns are decelerating the transition.

Rabobank's sustainability policies on the food and agribusiness sector

Rabobank's food and agribusiness loan portfolio (global) is €92.3 billion (21% of the total group loan) comprising Dutch farmers (85%), the Dutch food and agribusiness value chains (40%) and 2 million farmers in Africa, Asia and South America (Rabobank, 2015). Rabobank is participating in knowledge programmes for entrepreneurs and has established partnerships with various international organizations pertaining to the food and agribusiness sector. 'Banking for Food' is Rabobank's vision on global food security and how it describes its role. The 'Banking for Food' vision explains how to deal with issues regarding land and natural resource limitations against a mounting need for food at the global level (Rabobank, 2017c). The bank highlights four main sustainability concerns about food and agribusiness: availability, balanced nutrition, access and stability. Availability aspects cover solutions to improve food availability, covering innovations in sustainable production methods. Balance nutrition is about ways to stimulate a balanced nutrition by increasing awareness among consumers and improving the quality and safety of food. Access is about improving the availability of food by increasing the access to finance, knowledge and improving infrastructure and transfer/transport systems. Stability is a matter of creating a greater horizontal and vertical integration along the supply chain and removing the imbalance between market powers in the food supply chain (Rabobank, 2015).

The bank's food and agribusiness principles are formulated in terms of five guiding principles:

1   Safe and sufficient food production. This principle relates to the population growth forecasts and the necessity to increase the global food production capacity. Reasonable price and safety of the products are also indicated.
2   Responsible use of natural resources: issues such as soil degradation, erosion, soil and surface pollution, overfishing, clean water and preservation of natural areas are represented as factors to be considered.

3 Promotion of smart choices on the part of citizens and customers: the principle marks the importance of raising awareness among consumers about the importance of sustainable products.

4 Responsible animal husbandry and stock market: the principle is about the ethical and legal standards for animal rights.

5 Promoting social welfare: the importance of eliminating corruption, discrimination, human rights violations, forced labour, harmful child labour practices and poor working conditions in food and agribusiness chains (Rabobank, 2017d).

In addition to these five guiding principles, the bank has published various reports and informative articles on food and agribusiness which highlight the necessity of sustainable solutions and the bank's responsible attitude. Some sectors such as palm oil, seafood and coffee are prioritized; for them, sectoral objectives are set. However, these publications present limited solutions to initiate sustainable food and agribusiness. Issues such as (1) the role of the bank as a transformer; (2) the importance of sustainable supply chain management; (3) stakeholder engagement; (4) information sharing with the clients; and (5) networking are not fully embraced. In addition, forecasts and specific objectives on sustainable food and agribusiness credit portfolio seem to be missing. To address these fundamental gaps, questions regarding credit risk and rating models were directed to the interviewee.

According to the interviewee of Rabobank, the bank's general credit risk assessment is based on a 'know your client' policy, which has three main aspects: the client's financial strength, government rules and regulations and the bank's evaluation of the sector and/or the client. Especially, in the food and agribusiness sector, government regulations in The Netherlands are very strict while Rabobank has a proactive stance in complying with the legal requirements when considering financing any business. To a considerable extent, this already implies a substantial degree of sustainability. The bank promotes sustainability but does not oblige its clients to go along with it. Two main reasons were given for this approach. First, Rabobank has a long-term, deep-seated relationship with its clients, which has been based on voluntarism rather than coercion. However, it is stated that, when entire client groups develop a clear preference for particular forms of sustainability, the bank will be in a position to push harder or even will require an individual client to come along. Second, the legal requirements in The Netherlands are so strict that, normally, further admonitions are not considered necessary.

Furthermore, although the bank underlines the importance of innovative approaches in the food and agribusiness, this is not about new start-ups as the bank requires clients to have a financial track record of at least three years. In certain cases, however, the bank can help involve other financiers to arrive at a feasible (sustainable) investment. The bank's attempts at promoting crowdfunding can be given as an example of this effort. The bank utilizes its large network and credibility to bring investors and entrepreneurs together to provide funding for the innovative businesses. Nonetheless, it can be asserted that being a sustainable bank requires special financial support for entrepreneurs who aspire to aim for advances in sustainability.

It can also be stated that Rabobank does not have lending policies and risk assessment models specifically designed for sustainable financing, even though the sustainable framework of the bank and sector handbooks continuously emphasize the necessity to transform 'business as usual'. Here, the tension between conducting conventional banking policies and presenting sustainable business models to the clients is striking. According to the interviewee, especially in the food and agribusiness, Rabobank's client profile is mostly characterized by traditional family-owned enterprises which only reluctantly change their customary ways of doing business. Therefore, currently, the bank predominantly focuses on raising awareness about the importance of sustainable business practices and solutions.

According to publications of the bank, sustainable supply chain management is vital for the sustainable food and agribusiness. Raising awareness, establishing fair trading conditions, safe and sufficient food production and responsible use of natural resources are prominent issues. To heed these issues, the bank acts along with governments and NGOs wherever necessary. For example, currently, the Dutch government has launched an action plan for revitalizing pig farming (called *Vitalisering Varkenshouderij*); Rabobank plays a role in this action plan by providing access to relevant knowledge and networks (Rabobank, 2017d). Moreover, the bank participates in programmes with NGOs and the Dutch government to foster sustainable practices in the food and agribusiness sector. For example, the 'implementation agenda for the sustainable livestock farming sector' (*Uitvoeringsagenda Duurzame veehouderij*) aims to have a fully sustainable livestock sector by 2023; Rabobank plays an active role in helping farmers execute sustainable business practices (Rabobank, 2017d). During the interview, these contributions were also discussed. According to the interviewee, these policies are essential for the sustainable future of the sector. The bank plays a crucial role by being an advocate of these government policies while assisting in the transformation of the conventional way of doing business. To conclude, according to information from the interview and a review of available publications, it can be concluded that the bank's policies on sustainable food and agribusiness are primarily concentrated on raising awareness, dialogue and providing guidance for established clients who opt for sustainability. However, the bank will only take a strong role in promoting sustainability in projects and other business activities if it can rely on enabling government policies and a broad moral support among its clients.

In regard to supply chains, the interviewee stated that the bank does not have means and mechanisms to supervise the supply chain. As most of the food and agribusinesses have an international business structure, the bank is incapable of controlling and monitoring much of the business processes. However, interviewee of the monitor Fair Banking Guide opined that Rabobank could use its leverage to induce its clients to apply sustainable business models that include their supply chains. Today, the bank could do much more to promote sustainability in the food and agribusiness supply chains; its present policies in this area are rather weak; the roundtables with clients and NGOs were considered insufficient. For the interviewee of the monitor Fair Banking Guide, the bank should apply more result-oriented substantial policies. However, the bank claims that much remains invisible in the supply chains; it considers itself incapable of monitoring and correcting the things that go wrong in the supply chains. Here, the solution that is suggested to Rabobank is the certification of available standards. Rabobank should ask for certified products and services, which also cover the supply chain. In this way, environmentally harmful and socially undesirable business practices could be reduced.

As a conclusion, the bank's policies and practices in the food and agribusiness sector are in line with its overall position towards sustainability. In general, the vision and mission statements, which cover various aspects of sustainable development, reveal a responsible attitude. Nonetheless, current banking operations are still frequently executed by means of conventional methods. The bank's practices can be seen as rather active (in terms of the Tulder model) but not across the board. One practice-based indicator is the total sustainable credit portfolio of the bank (€ 18.8 billion in 2016), which is substantial in absolute terms. However, this figure is less than 5% of the whole portfolio. Also, there is room for a much more systematic approach towards sustainability in support of a constantly enhanced sustainability performance in ordinary banking activities. In terms of the Tulder model, Rabobank can be classified as a bank that is largely reactive but, on several issues, it conducts a range of active policies. These could be the starting point of a further development towards a fully proactive, sustainable bank.

Conclusion

When considering the long-standing history of Rabobank, its deep-rooted relationship with its clients and the competitive structure of the banking sector, it seems unrealistic to expect an unfettered transformation. The case study of Rabobank shows that a commercial bank's full transformation can be promulgated but not exacted from the clients without a broader societal and governmental support. However, the sectoral policies and practices of a commercial bank can vary. In this regard, it is necessary to be flexible in one's analyses, considering the structural and cultural factors of a particular sector and the opportunities these may offer (Table 6.5).

On one hand, running long-term relationships with one's clients can be seen as a major business asset for Rabobank, but, on the other hand, these can be a powerful handicap, frustrating a proactive sustainable approach. In any case, it seems important to make a distinction between existing and new clients. Accordingly, the bank can apply its sustainability standards to its lending practices and other sustainability-related issues. By doing so, the bank's involvement with socially and ecologically harmful businesses would be confined. Certainly, the bank must keep attempting to increase existing clients' sustainability awareness and

TABLE 6.5 Areas of sustainability performance analysis

	Areas of Sustainability Performance Analysis	
	Existing Clients	New Clients
General Strategy and Practices		
Sectoral-based Strategy and Practices		

performance. Despite having a dual approach towards clients, contributing sustainable development proactively and fully should remain as the ultimate objective of the bank.

The second variable of the template (see Table 6.5) highlights the difference between the bank's general and sectoral base strategy and practices. This classification is derived from the its diversified attitude towards particular sectors and sub-sectors (depending on economic and technological opportunities). In this way, the transition of the bank towards sustainability can gain momentum.

To sum up, by differentiating between old and new clients and differentiating between sectors, Rabobank could overcome certain structural and operational obstacles by executing an active or even proactive approach to sustainable development.

The mystery of sustainable banking seems to lie in motivating and inducing ordinary people to go beyond what the conventional order seems to be happy with. Leadership, responsibility, consistent policies and collaboration with the right partners are its ingredients.

The Tulder model applied to the banking sector has proved to be a valuable tool in analysing existing banking practices and in suggesting new roads to sustainability. However, this cannot be a matter of a rigid exercise; it requires a feel for the circumstances and a realization that transformation is far from easy.

References

ABN AMRO Bank N.V (2016) Sustainable Risk Policy. Available at: www.abnAMRO. com/en/images/040_Sustainabe_banking/Links_en_documenten/Documenten/Beleid_- _Sustainability_Risk_Policy_EN.pdf: ABN AMRO Bank N.V.

Anderson, D.R. (2001) *Misguided Virtue. False notions of corporate social responsibility.* Wellington: New Zealand Business RoundTable.

Arnsperger, C. (2014) On the politics of social and sustainable banking, *Global Social Policy*, 14(2), pp. 279–81.

Aziakpono, M., Bauer, R. and Kleimeier, S. (2014) Financial globalisation and sustainable finance: Implications for policy and practice, *Journal of Banking and Finance*, pp. 137–138.

BBMG, G.A.S. (2012) Re: Thinking Consumption Customers and the Future of Sustainability', Available at: www.globescan.com/component/edocman/?task=docu ment.viewdoc&id=51&Itemid=0 (accessed: 5 Jan. 2018).

Burchell, J. and Cook, J. (2006) It's good to talk? Examining attitudes towards corporate social responsibility dialogue and engagement processes, *Business Ethics: A European Review*, 15(2), pp. 154–171.

Carroll, A.B. (1979) A three-dimensional conceptual model of corporate performance, *Academic Management*, 184, pp. 497–505.

Carroll, A.B. (1993) *Business and Society: Ethics, Sustainability, and Stakeholder Management.* 2nd edition. Cincinatti, OH: South-Western Publishing.

Carroll, A.B. and Buchholtz, A.K. (2008) *Business & Society, Ethics and Stakeholder Management.* 7th edition. Stamford, CT: Cengage Learning.

Chichilnisky, G. and Heal, G. (2000) *Environmental Markets: Equity and Efficiency.* New York: Columbia University Press.

Davis, K. and Bloomstorm, R. L. (1975). *Business and Society: Environment and Responsibility.* 3rd edition. New York: McGraw-Hill.

Dietz, S. and Neumayer, E. (2007) Weak and strong sustainability in the SEEA: Concepts and measurements', *Ecological Economics*, 61, pp. 617–626.

Donaldson, T. and Preston, L.E. (1995) The stakeholder theory of the corporation: Concepts, evidence, and implications', *Academy of Management*, 20(1), pp. 65–91.

Eccles, R.G. and Serafeim, G. (2013) The performance frontier: Innovating for a sustainable strategy, *Harvard Business Review*, 91(5), pp. 50–60.

EerlijkeBankwijzer (2014) 5 Years of the Dutch Fair Bank Guide. Available at: http://eerlijkegeldwijzer.nl/media/60485/5-years-of-the-dutch-fair-bank-guide-januari-2015 compressed.pdf (accessed 5 January 2018).

EerlijkeGeldwijzer (2015) Theory of change – Eerlijke Geldwijzer. Available at: http://eerlijkegeldwijzer.nl/media/60596/150319-theory-of-change-eerlijke-geldwijzer.pdf (accessed 5 January 2018).

Elkington, J. (1998) *Cannibals with Forks: Triple Bottom Line in 21st Century Business.* Gabriola Island, Canada: New Society Publishers.

Freeman, E.R. (1984) *Strategic Management: A stakeholder approach.* London: Pitman.

Freeman, E.R., Harrison, J.S., Wicks, A.C., Parmar, B.L., and De Colle, S. (2010) *Stakeholder Theory, The State of the Art.* New York: Cambridge University Press.

Freeman, E.R., Wicks, A.C. and Pamar, B.L. (2004) Stakeholder theory and 'The Corporate Objective Revisited', *Organized Science*, 15(3), pp. 364–369.

Frynas, J. G. and Yamahaki, C. (2016) Corporate social responsibility: Review and roadmap of theoretical perspectives, *Business Ethics: A European Review*, 25(3), pp. 258–285.

Gilbert, D.U. and Rasche, A. (2008) Opportunities and problems of standardized ethics initiatives – A stakeholder theory perspective, *Journal of Business Ethics*, 82(3), pp. 755–773.

Goel, P. (2010) Triple bottom line reporting: An analytical approach for corporate sustainability, *Journal of Finance and Management*, 1(1), pp. 27–42.

Goss, A. and Roberts, G.S. (2011) The impact of corporate social responsibility on the cost of bank loans, *Journal of Banking and Finance*, 35, pp. 1794–1810.

Haupt, U. and Henrich, U. (2004) *Sectoral Policy Paper on Financial System Development*, Bonn: Federal Ministry for Economic Cooperation and Development.

Herring, R.J. and Santomero, A.M. (1996) *The Role of the Financial Sector in Economic Performance.* Working Paper, Philadelphia, PA: The Wharton School (University of Pennsylvania), pp. 1–90.

Hopwood, B., Mellor, M., and O'Brien, G. (2005) Sustainable development: Mapping different approaches, *Sustainable Development*, 13, pp. 38–52.

Jeucken, M.H. and Bouma, J.J. (1999) The changing environment of banks, *Greener Management International*, 27, pp. 27–35.

Laugel, J.-F. and Laszlo, C. (2009) Financial crisis: The opportunity for sustainable value creation in banking and insurance, *Journal of Corporate Citizenship*, 35, pp. 24–38.

Logsdon, J.M. and Wood, D.J. (2002) Business citizenship: From domestic to global level of analysis, *Business Ethics Quarterly*, 12(2), pp. 155–187.

Lubin, D.A. and Esty, D.C. (2010) The sustainability imperative, *Harvard Business Review*, 88(5), pp. 42–50.

McWilliams, A., Parkhankangas, A., Coupet, J., Welch, E., Barnum, D. (2014) Strategic decision making for the triple bottom line, *Business Strategy and the Environment*, 25(3), pp. 193–204.

Mitchell, R.K., Agle, B.R., and Wood, D.J. (1997) Toward a theory of stakeholder identification and salience: Defining the principles of who and what really counts, *The Academy Management Review*, 22(4), pp. 853–886.

Parmigiani, A., Klassen, R.D., and Russo, M.V. (2011) Efficiency meets accountability: Performance implications of supply chain configuration, control, and capabilities, *Journal of Operations Management*, 29, pp. 212–223.

Perez, A. and Rodriguez del Bosque, I. (2014) Sustainable development and stakeholders: A renewal proposal for the implementation and measurement of sustainability in hospitality companies, *Knowledge and Process Management*, 21(3), pp. 198–205.

Polonskaya, J. and Babenko, M. (2012) *Best Practice Guide on Sustainable Finance: A Practical Toolkit for Russian Financial Sector*. WWF Sustainable Finance Programme Report. Berlin: WWF.

Rabobank (2015a) Facts & Figures Banking for Food, Utrecht. Available at: www.rabobank.com/en/images/2015%2003%20Banking%20for%20Food_facts%20%20figures%20Feb%202015.pdf (accessed 5 January 2018).

Rabobank (2015b) How to Guarantee -sustainably- Food Security? Utrecht. Available at: www.rabobank.com/en/images/How-to-guarantee.pdf (accessed 5 January 2018).

Rabobank (2016a) Infographic – Impact 2016, Utrecht. Available at: www.rabobank.com/en/images/impact-2016-infographic-en.pdf (accessed: 5 January 2018).

Rabobank (2016b) Infographic Rabobank 2016, Utrecht. Available at: www.rabobank.com/en/images/02-infographic-rabobank-in-2016-eng.pdf (accessed: 5 January 2018).

Rabobank (2016c) Sustainability Policy Framework, Utrecht. Available at: www.rabobank.com/en/images/sustainability-policy-framework.pdf (accessed: 5 January 2018).

Rabobank (2017a) Annual Report 2016, Utrecht. Available at: www.rabobank.com/en/images/annual-report-2016.pdf (accessed: 5 January 2018).

Rabobank (2017b) Banking for Food: Vision on Global Food Security and The Role of Rabobank. Available at: www.rabobank.com/en/about-rabobank/food-agribusiness/vision-banking-for-food/index.html (accessed: 5 January 2018).

Rabobank (2017c) Rabo's Food & Agribusiness Principles. Available at: www.rabobank.com/en/about-rabobank/food-agribusiness/principles/index.html (accessed 5 January 2018).

Rabobank (2017d) Vision and policy: Sustainably successful together. Available at: www.rabobank.com/en/about-rabobank/in-society/sustainability/vision-and-policy/vision-sustainably-successful-together.html (accessed: 5 January 2018).

Rabobank Group (2016) Annual Report 2015. Available at: www.rabobank.com/en/images/rabobank-annual-report-2015.pdf: Rabobank (accessed 5 January 2018).

Ramnarian, T.D. and Pillay, M.T. (2016) Designing sustainable services: The case of Mauritian banks, *Social and Behavioral Sciences*, 224, pp. 483–490.

Schaltegger, S., Hansen, E.G., and Lüdeke, F. (2015) Business models for sustainability. Origins, present research, and future avenues, *Organization & Environment*, 29(1), pp. 3–10.

Schaltegger, S. and Wagner, M. (2011) Sustainable entrepreneurship and sustainability innovation: categories and interactions, *Business Strategy and the Environment*, 20(4), pp. 222–237.

Scholtens, B. (2006) Finance as a driver of corporate social responsibility, *Journal of Business Ethics*, 68(1), pp. 19–33.

Scholtens, B. (2007) Financial and social performance of socially responsible investment in The Netherlands, *Corporate Governance: An International Review*, 15(6), pp. 1090–1105.

Scholtens, B. and Van Wensveen, D. (2003) *The Theory of Financial Intermediation: An Essay on What It Does (not) Explain.* SUERF Studies, Volume 1. Vienna: The European Money and Finance Forum.

Schulz, S. and Flanigan, R.L. (2016) Developing competitive advantage Using the triple bottom line: A conceptual framework, *Journal of Business & Industrial Marketing*, 31(4), pp. 449–458.

Slaper, T.F. and Hall, T.J. (2011) The triple bottom line: What is it and how does it work?', *Indiana Business Review*, 86(1), pp. 4–8.

Steurer, R., Langer, M.E., Konrad, A., and Martinuzzi, A. (2005) Corporations, stakeholders and sustainable development I: A theoretical exploration of business-society relations, *Journal of Business Ethics*, 61(3), pp. 263–281.

Tulder, R.V. and Fortainer, F. (2009) Business and sustainable development: From passive involvement to active partnership. In Kremer, M., van Lieshout, D., and Went, R. *Doing Good or Doing Better. Developing Policies in a Globalising World.* The Hague/ Amsterdam: Amsterdam University Press, pp. 211–228.

Tulder, R.V., Tilburg, R.V., Francken, M., and Rosa, A.D. (2014) *Managing the Transition to a Sustainable Enterprise. Lessons from frontrunner companies.* Abingdon, UK: Routledge.

UNEP (2015) The Financial System We Need, Aligning the Financial System with Sustainable Development. Geneva: UNEP.

UNEP Finance Initiative (2007) Insuring for Sustainability: Why and how the leaders are doing it?, Paris: UNEP Finance Initiative.

UNEP Finance Initiative (2015) Banking and Sustainability Time for Convergence. Paris: UNEP Finance Initiative.

United Nations Global Impact (2016) Short-Termism in Financial Markets. Available at www.unglobalcompact.org/take-action/action/long-term (accessed: 5 January 2018).

United Nations (1987) Our Common Future, Brundtland Report. Available at: www.un.org/en/ga/search/view_doc.asp?symbol=A/RES/42/187 (accessed: 5 January 2018).

Vigano, F. and Nicolai, D. (2009) CSR in the European banking sector: Evidence from a survey. In: Barth, R. and. Wolff, F. (eds) *Corporate Social Responsibility in Europe: Rhetoric and realities.* Cheltenham, UK: Edward Elgar, pp. 95–109.

Walter, S.R. (2016) Circular economy, *Nature*, 531, pp. 435–438.

Weber, O. (2005) Sustainability benchmarking of European banks and financial service organizations, *Corporate Social Responsibility Environment Management*, 12, pp. 73–87.

Weber, O. (2012) Sustainable banking – history and current developments. Available at: www.researchgate.net/profile/Olaf_Weber2/publication/254932876_Sustainable_Banking__History_and_Current_Developments/links/00463530f9abcb09a1000000.pdf (accessed: 5 January 2018).

Weber, O. (2013) Sustainable banking – history and current developments. EMES-SOCENT Conference Selected Papers. Available at: www.emes.net (accessed 5 January 2018).

Weber, O. and Feltmate, B. (2016) *Sustainable Banking, Managing the Social and Environmental Impact of Financial Institutions.* Toronto: University of Toronto Press.

Wolters, T. (2013) *Sustainable Value Creation as a Challenge to Managers and Controllers.* Apeldoorn: Wittenborg University Press.

Wright, C. (2012) Global banks, the environment, and human rights: The impact of the Equator Principles on lending policies and practices, *Global Environmental Politics*, 12(1), pp. 56–77.

Yin, R.K. (1994) *Case Study Research Design and Methods*. Second 2nd. London: SAGE.

7

ISLAMIC FINANCE AND SUSTAINABILITY

Muhammad Ashfaq and Teun Wolters

The Islamic view on sustainability[1]

Different studies show that in Islamic teaching various values come to the fore that can easily be related to, what we call today, sustainability. In general, there is an emphasis on individual choice in striking a balance between responsible behaviour and piety.

That balance derives from recognizing that human beings have two missions to fulfil: (1) the mission as a servant of Allah and (2) the mission as vicegerent or steward of Allah. In the first mission a person is responsible to God as His servant, while the second mission is about a person's relationship with creation.[2] Both missions, and relationships at the same time, have an equal weight in what makes behaviour virtuous. There is to be a balance, that is, things ought to be done in a proportionate manner, avoiding extremes.[3]

Islam has a broad conception of worship. This requires a balance between the demands of this world and the demands of the afterlife (Chapra, 1992). The importance and special nature of worship are fundamental elements of Islam. It comprises the five well-known components of confession, prayer, fasting during Ramadan, charity and the pilgrimage to Mecca.

However, worship is not limited to these points; Muslims need to show good behaviour in all aspects of their daily life, even in their work and business life. Many passages in the Koran encourage economic activity as every individual is required to work (Williams and Zinkin, 2010, 520).

All forms of productive work can be seen as an act of worship if the material gain accruing from it leads to social justice and spiritual enhancement. However, daily actions can only be seen as part of worship if (1) these are undertaken wholeheartedly for the sake of Allah (and not for other reasons such as the love for money); (2) the actions should accord with the *Sariah*; and (3) they must not cause a Muslim to neglect existing obligations.

Islam recognizes that responsibility in business goes beyond what is done with the profits made from business activities. Business is encouraged because it provides sustenance. However, certain types of business are prohibited, especially concerning those products and services that put human health at risks (alcohol, tobacco, armaments and gambling).

The prohibition on usury and interest is also well known. As most scholars in this area see it, usury and interest are forbidden because they create profits without work and there is no sharing of risk between the lender and the borrower (Williams and Zinkin, 2010, 523). Speculation is not acceptable. A buyer must be fully informed of a product that he/she is considering buying, while the seller should have it under his/her control.

Taken together, it seems that Islam has its own basis for a society: it prefers an equity-based, risk-sharing and stake-taking economic system to a debt-based system (Ahmad, 2003). This conclusion also puts Islamic finance in perspective, as will be discussed later.

Williams and Zinkin (2010) compare the tenets of Islam to the UN's Global Compact code of conduct. The Global Compact is a voluntary process aiming at bringing a set of universal principles of responsible business into mainstream

TABLE 7.1 The ten principles of the UN Global Compact

Human Rights	
Principle 1	Businesses should support and respect the protection of internationally proclaimed human rights.
Principe 2	Ensure that businesses are not complicit in human rights abuses.

Labour	
Principle 3	Businesses should uphold the freedom of association and the effective recognition of the right to collective bargaining
Principle 4	The elimination of all forms of forced and compulsory labour
Principle 5	The effective abolition of child labour
Principle 6	The elimination of discrimination in respect of employment and occupation

Environment	
Principle 7	Businesses should support a precautionary approach to environmental challenges
Principle 8	Undertake initiatives to promote greater environmental responsibility
Principle 9	Encourage the development and diffusion of environmentally friendly technologies

Anti-corruption	
Principle 10	Businesses should work against corruption in all its forms, including extortion and bribery

activities undertaken by companies undertake all over the world. It also strives to be a catalyst for initiatives in support of wider UN social and environmental goals. Global Compact is intended to be a coherent framework for sustainable business. The ten UN Global Compact principles are reflected in Table 7.1.

Islam and the UN Global Compact

The 10 principles launched by Global Compact are rooted in the 1948 Universal Declaration of Human Rights and cover a variety of factors relating to the dignity, freedoms and protection of the individual. Williams and Zinkin (2010) conclude that Muslim businesses should have no problem complying with the UN Global Compact as the Koran and the Shariah often go further than required by Global Compact.

Equal treatment

The right to equal treatment is inherent in Islam; it recognizes the brotherhood of man irrespective of race, colour or nationality. Islam makes it clear that non-Muslims are to be treated respectfully, while there should not be compulsion in regard to religion. At least according to early Islam, women have equal rights to men. According to Beekun and Badawi (2005: 137), normative Islam rejects sexism, both in business and in other areas of life. The Koran's only basis for superiority is piety and righteousness, not gender (Williams and Zinkin, 2010, 524). The right to equal treatment extends to equality before the law. Unfortunately, Islam's normative teachings are inconsistently followed in the Muslim world or often set aside.

Life and security

Islam explicitly confesses the protection of the right to life. Serious crimes need to be brought before a competent court.

Environmental sustainability according to Islam

Islam expects the believer to respect natural resources (see Box 7.1). These should only be used to provide necessities – that means that a luxurious life with extravagance should be avoided. Jusoff and Alam (2011) claim that the Prophet Mohammad was 'an environmentalist avant la lettre'; he had a profound respect for fauna and flora; he believed that not only animals, but also land, forests and watercourses, should have rights.

Environmental issues are deeply steeped in the moral consciousness of a culture. In fact, they are part and parcel of a religious view of the world. Human–environment interactions exist with dynamic cultural, spatial and temporal contexts. Therefore, it is critical that our natural resource management strategies incorporate

BOX 7.1 EIGHT SELECTED ISLAMIC PRINCIPLES AND GUIDELINES ON SUSTAINABILITY

Adl (Justice): governing human relationships and other living creatures

Mizan (Balance): governing not only human, social and economic relationships but also the environment, especially in ensuring the equilibrium of nature, use of resources and life cycle of all species.

Wasat (Moderation): choosing the middle path in economic planning, social conduct, scientific pursuits, ideological views, material, water and energy consumption.

Rahmah (Mercy): governing all aspects of human relationships and treatment of all living animals, plants and insects including micro-organisms

Amanah (Trustworthiness and custodianship): Humankind is considered to be a trustee appointed by the Creator, for all earth's assets.

Taharah (Spiritual purity and physical cleanliness): generating contented individuals through spiritual purity, conscious of the presence of his/her Creator that would result in a balanced society, living in harmony with the environment; cleanliness that would generate a healthy lifestyle; society devoid of air and water pollution, as well as generating a clean economy devoid of usury and deceitful marketing techniques and business transactions.

Haq (Truthfulness and rights): truthfulness in all dealings that recognizes the respective rights of others (humans, animals and plants).

Ilm Nafi' (Usefulness of knowledge and science): knowledge, whether theological, scientific or technological, must be beneficial to others (individuals and society) including future generations.

Source: Matali, Z. *Sustainability in Islam* (earthcharter.org, retrieved 20 October 2017)

elements of local cultures and religions (Jusoff and Alam, 2011). However, moderation is an overarching principle; human relationships in Islam have to be based on justice and kindness, and not on economic gain. From there, it can be said that the prevention of pollution and avoiding a wasteful use of natural resources are binding obligations (Jusoff and Alam, 2011).

Islamic finance

Islamic finance can be directly situated within the above theological vision of Islam on sustainability, although it is important to realize that the principles discussed need to be adequately linked with the contemporary sustainability issues.

Islam is a complete code of life and offers solutions to a wide range of problems arising from personal, social, political and economic situations (Mohamad *et al.*, 2014: 208). In the view of Malkawi (2014, 41), in Islam there are certain rules and ethical norms that provide guiding principles for conducting business and economic activities.

Olaoye *et al.* (2013, 24) cite the definition of Islamic banking presented by the prominent Islamic scholar Siddiqi as:

> Islamic banking as the business of financial intermediation, mobilizing savings from the public on the basis of partnership and profit and advancing capital to entrepreneurs on the same basis. Islamic banking, therefore, is a system of banking that complies with the principles of Shari'ah (Islamic law) and its practical application through the development of Islamic economics.

The Islamic financial system prohibits charging interest (Riba) on loans (Askari *et al.*, 2012, 50; Daly and Frikha, 2015, 1). Both the borrower and the lender must share risk in any transaction. Both the provider of capital and the entrepreneur share the businesses profit and loss. Speculative behaviour (Gharar) and gambling (Maysir) are not allowed in Islamic Shari'ah. Due to the involvement of speculation and uncertainty, Islamic financial practices also restrict the use of derivatives (financial instruments that derive their price from one or more underlying assets) (Al-Amine, 2013b, 332). Money is seen as potential capital and thus only takes the form of actual capital when it is used in a productive capacity. Sinful products, which are banned by Quran, including such products as pork, alcohol and prostitution, cannot be financed (Alrifai, 2015, 117–21).

Islamic finance involves forms of banking (savings and credits) and insurance.

El-Hawary *et al.* (2004, 5) mention four principles comprising the foundation of modern Islamic banking:

(1) Risk-sharing: the terms of financial transactions need to reflect a symmetrical risk/return distribution among each participant to the transaction.
(2) Materiality: all financial transactions must have material finality, i.e., be linked to underlying assets. It means that financial assets such as options and derivatives without the backing of real assets are not acceptable.
(3) No exploitation: the transaction should be fair and transparent and no party should have undue influence and advantage.
(4) No financing of sinful activities: financing given only to permissible activities in the views of the Quran and Sunnah is allowed. Therefore, financing of activities such as those involving alcohol, pork and gambling is strictly prohibited.

Tools of Islamic finance

The Islamic financial system operates as a financial intermediary between those who have excess resources (primarily households) and those who require them for investment purposes (primarily businesses). This important role of financial intermediation is based on the use of a number of financial instruments, contracts and tools that have been developed over time. Islam places a great deal of emphasis on trade and encourages all participants to write down their business dealings in the form of contracts. Islamic finance uses a set of financial instruments, contracts and tools for 'real economic activities, financing, intermediation and social welfare' (Greuning and Iqbal (2008, 17). Below are a number of such instruments and tools.

Musharakah *(equity partnership, joint venture)*

Musharakah is a kind of partnership where two or more people contribute labour and financial resources on the basis of profit and loss sharing (PSL) (Iqbal and Molyneux, 2005, 20).

Mudarabah *(investment partnership)*

Mudarabah is a type of equity-based partnership contract that is also known as trust financing, and is one of the core profit- and loss-sharing contracts in Islamic finance. It is very similar to the conventional limited partnership contract in which one partner contributes capital and the other contributes management skills, while profits are shared based on an agreed upon ratio (Iqbal and Mirakhor, 2011, 90; Kettell, 2011, 62).[4]

Murabahah *(cost-plus sale/mark-up financing)*

Murabahah is one of the most widely used Islamic finance instrument (Alrifai, 2015, 128). This instrument is used for trade financing. *Murabahah* is also known as cost-plus sale and mark-up financing.

Bay' Salam *(forward sale)*

Bay' Salam is an advanced-sale type of contract, commonly known as a forward sale contract; it is one of the most exceptional instruments in Islamic finance. Under this type of contract, a seller can sell a commodity against advance payment without transferring possession and ownership of that commodity to the buyer. The Prophet Muhammad allowed this contract to enable farmers to receive money to purchase necessary inputs of production for farming (Iqbal and Mirakhor, 2011, 80). *Bay' Salam* was also used to finance import and export conducted by the merchants of Arabia, who were allowed to sell their commodities in advance to obtain finances for their businesses. A *Bay' Salam* contract looks similar to a conventional forward-sale contract. However, it differs in terms of payment method because in the former a buyer pays the full price at the time of entering into an agreement and in advance of receiving the goods; this is the main difference between *Bay' Salam* and the conventional forward-sale contract.

Ijarah *(leasing)*

The literal meaning of the word *Ijarah* is to give something on rent; leasing in conventional finance is similar to the Islamic concept of *Ijarah* (Lewis, 2009, 17; Ernst *et al.*, 2013, 38). *Ijarah* is the transfer of rights of use (usufruct) imbedded in a specific property or item to another person in exchange for rent payments made for a specific amount of time. *Ijarah* is not a financing instrument but rather a transfer of usufruct for an asset. Gradually, Islamic banks have begun to use it as one of their core instruments.

Islamic finance and environmental sustainability

The Islamic finance industry has been active in contributing to attaining the goal of global environmental sustainability. A number of Socially Responsible Investments (SRIs) or green Sukuk have been introduced and launched at the global level to support the financing of those projects that are environmental friendly. For example, in 2012, the Climate Bonds Initiative (CBI) with the collaboration of the Clean Energy Business Council of the Middle East and North Africa (MENA) and Dubai-based Gulf Bond & Association formed and started the Green Sukuk Working Group to develop and endorse the concept of green Sukuk which matches the low-carbon criterion. Likewise, in 2014, the Securities Commission SC Malaysia re-edited its Sukuk guideline by including the new requisites for launching the SRI Sukuk. The revised Sukuk guideline elucidates that the revenue from the SRI Sukuk can also be utilized to preserve natural and ecological resources, save on energy usage, enhance renewable energy usage and diminish greenhouse gas emissions.

Here follow a few interesting examples.

- In 2012, two Australian solar companies (Solar Guys International and Mitabu) generated funds in Indonesia amounting to a value of USD 100 million for a 50 MW photovoltaic project based on green Sukuk. The project was basically designed in Malaysia and was fully funded under a Power Purchase Agreement.
- The Islamic Development Bank (IDB) has funded (involving USD 180 million) pilot projects on clean energy in its 56 member countries around the globe.
- In 2014, the International Finance Facility for Immunisation (IFFIm), based in the UK, launched SRI Sukuk, based on *Murabahah*, having a value of USD 500 million. The money became available for the immunization of children in the poorest countries of the world via Gavi, the Vaccine Alliance. Thanks to this campaign, tens of millions of children could be protected against vaccine-preventable disease (Malaysian Islamic Finance Centre (MIFC), 2016).

Islamic micro-finance

Islamic micro-finance is a market niche providing financial mediation with emphasis on ethical and moral values in business according to Islamic teaching. It addresses deprivations and vulnerabilities of the poor; it offers a variety of products such as micro-credit, micro-savings and micro-insurance in a way that supports social cohesion of small groups, communities, institutions or societies (Tavanti, 2013).

Usman and Tasmin (2016) carried out many studies related to Islamic micro-finance. They concluded that Islamic micro-finance is a basic tool for reducing poverty and enhancing the livelihoods of so many people by providing better-conditioned houses. It is also known as an instrument in empowering poor women to develop and establish micro-enterprises. Ideally, it also promotes entrepreneurial education, self-reliance, skills building, assets accumulation and communal services. Usman and Tasmin (2016) indicate that Islamic micro-finance is conducive to the achievement of the UN's Strategic Development Goals.

Charity-based funding

Wasq and *zadat* (forms of charity) can also be used to enhance the resilience of the needy. For example, Diwan al Zakat, an institution based in Sudan, has started to provide loans to farmers at the initial period of the agricultural season so that they will be able to buy most essential and basic inputs; the loans thus provided are to be repaid after the time of harvest. Results show that this implemented policy has enhanced the capability of farms in terms of productivity; also, the collection of *zadat* increased from farmers, amounting to 74.4% of the loans provided. Another good method for reducing susceptibility is the use of *waqf* and *zakat* funds to pay the monthly *takaful* premiums to create security against various specified risks (Ahmed *et al.*, 2015).

A research study in the context of South East Asian countries (Malaysia, Indonesia, Pakistan) found that obligatory alms, including *zakat*, are dependable and viable. These can be a key tool in collecting funds for particular organizations; it was shown that this form of fundraising improved their trustworthiness by scaling up their good governance, integrity and transparency (IRTI, 2015).

To enhance and accelerate the performance of *waqf* and *zakat* in achieving social development, it is necessary to address two major problems and issues: (1) augmenting the assets on which *zakat* can be collected by adapting the definition and scope of wealth to the needs of modern times; and (2) the need for a multifaceted scheme to reformulate the scope of the *waqf* sector along these lines. Some countries have already adopted this approach to revive the *waqf* sector, such as the Sudanese Awqaf Authority, which has generated funds through donations as a new *waqf* and has formed an investment/construction department to enhance existing *waqf* assets by making them efficient and more productive (Ahmed *et al.*, 2015).

Possible regulation of Islamic finance in Europe

Globalization has changed the ethnic composition of many countries, particularly in Western Europe with its large Muslim minorities. Being in a predominantly foreign social environment leads to concentration on issues such as a shared faith, language or cultural background. Often emigration leads to an increasing interest in religion and culture, as these are frequently seen as linkages with the homeland, even if the emigration concerned took place several generations earlier. This has led to an increased interest in financial products compatible with the teaching of Islam (Kirchner, 2011).

The European Convention on Human Rights (ECHR), Article 9, suggests that, although there are limitations in the interests of public safety, freedom of religion is by no means limited to the freedom of worship. To be effective, freedom of religion includes the right to live one's life in line with the rules and demands of one's religion. Article 9 of the ECHR also protects the observance of religious customs, which include not only rituals but also everyday life (Kirchner, 2011). However, many instruments of Islamic finance have been created fairly recently. Even if they may have an old religious basis, these new and diverse financial instruments can hardly be called a tradition. However, even if Islamic finance is not (yet) a custom, for a number of people it can be a 'practice of one's faith' according to Article 18 of the Universal Declaration of Human Rights (Kirchner, 2011).

Kirchner (2011) suggests improving the regulation of Islamic finance in Europe, as this could prevent a repeat of the large-scale fraud suffered by many investors in the 1990s. This could also be in the interest of the providers of Islamic financial services, as it improves the quality of the financial products and makes them more acceptable to a broader group of investors.

To allow for Islamic finance to be regulated by law means treating it as any other financial service. For that matter, the rules of Islamic finance would not be applied qua religion but as an approach that is not (necessarily) at odds with European legal principles.

One goal of regulating financial services is the protection of the customer. Within this context, regulators need to familiarize themselves with Islamic finance. Traditionally, three types of professionals are needed to set up new Islamic finance instruments: finance experts, shariah jurists and secular lawyers. The secular legislature has to protect investors, while shariah jurists must ensure that new financial products are truly shariah compliant to reduce the risk of abuse of the label 'Islamic' and fraud to the detriment of faithful Muslim investors. Shariah boards could be helpful in arbitrating financial disputes within the confines of Islamic finance. The state's regulatory work could then be limited to secular laws that do not differentiate according to religion (Kirchner, 2011).

Such a framework of regulation and freedom could also serve the purposes of sustainable development provided Islamic finance takes the principles upon which it rests seriously and considers the modern world's needs for breaking away from 'business as usual'.

A critical perspective

As explained, Islamic finance is largely determined by prohibitions regarding the purpose and structure of financing from the Koran and the tradition of the prophet Mohammad (in particular, no interest, no excessive uncertainty and no gambling). Forbidden is the consumption of alcohol and pork. The scope of such prohibitions extends to the production, financing and use of tobacco as it is a threat to health (Hayat and Malik, 2014). In the literature on Islamic finance there are all types of interpretive discussions which at times seem to be rather legalistic, representing a debate that is purely internal to Islamic thinking. Moderation and the avoidance of excessive uncertainty (and the risks involved) as akin to Islamic finance may be helpful in avoiding wasteful and pointless consumption; substantive action in support of sustainability would require further extensions of the classic Islamic finance principles to the modern world. The latter require not only interpretive skills but equally a sense of urgency and a kind of moral responsibility that urges financiers and producers to take sustainability to heart.

An important issue is the execution of screening processes to avoid the financing of companies whose primary funds are in conflict with Islamic jurisprudence. Such companies include those which operate in the 'sin' industries. After the initial screening based on a company's primary business, further exclusions are applied, in particular, to avoid business with a high level of impermissible interest income (usually 5%) or interest-bearing debts (this information is based on Hayet and Malik, 2014; this reference can be consulted for further details).

As Hayet and Malik (2014) explain, the above screening process is generally negative; unlike responsible investment, Islamic investing is not known to actively use engagement with companies as a strategy. They also claim that screening methods in Islamic finance, like many other aspects of Islamic finance, should be regarded as 'works in progress'; most companies are not Shariah compliant as an objective; if they are, it is by chance rather than by design.

Demuth (2014) interviewed Muslim representatives in Germany. She comes to the conclusion that Islamic finance is subject to scepticism in the various Muslim communities. Although Islamic finance is a form of ethical banking (or responsible investment) by its very nature, it is inclined to take conventional finance as a yardstick rather than the field of ethical banking and investment. Thereby, form seems to take precedence over substance. Demuth (2014) suggests that responsible investment (or SRI) can be informative to Islamic finance in terms of how to add substance to their activities. She mentions the religious roots of SRI; originally, it was driven by faith-based actors and other civil society organizations.

Generally, every Islamic financial institution has its shariah board, which consists of a number of scholars trained in a range of subjects such as economics, finance, legal studies or even arts (Ünal, 2013). The way these decisions are made is usually not made public in any great detail, whereas in responsible investment, transparency on how decisions were made is a major feature.

Beyond negative screening, responsible investment has a broader set of methods that can be used to support the ethical content of investment decisions (Demuth, 2014): negative screening (avoiding controversial business); positive screening (exemplary behaviour in areas such as social engagement, environmental protection and corporate governance); shareholder activism (using shareholder rights to confront their invested companies with critical issues and fighting for positive ethical behaviour from within); impact investment (generating a measurable positive social impact by supporting certain investments); and integration (promoting the integration of environmental, social and governance (ESG) criteria into the mainstream financial industry).

These methods are mainly used for the screening of publicly traded shares or bonds. However, besides screening, impact investment and integration can also be used in lending or project finance. Therefore, the various methods of responsible investment could also play a role in Islamic finance (Demuth, 2014). This broadening of methods would help to make the decisions within the framework of Islamic finance more transparent and understandable to a wider audience. Most of all, it could make Islamic finance a more effective supporter of sustainability in line with what it promises to be.

Notes

1 This section is largely based on Williams and Zinkin (2010).
2 This concept of stewardship seems to be close to that of Protestantism in Christianity, which has had an influence on business ethics in Europe. There seems to be a close affinity between Islamic concepts of individual responsibility and accountability to God and those of Protestantism (Williams and Zinkin, 2010; 529–30).
3 Also, see Jusnaidi (2015), who works out the three dimensions of sustainability (i.e. social, environmental and economic) based on Islamic thinking.
4 Nienhaus (2015: 161) argues that the *Mudarabah* contract is a rare Islamic financial contract where Islamic banks practise profit-and-loss sharing through deposit investment accounts of customers. However, in reality, use of it for this purpose also deviates from the genuine model of *Mudarabah* financing in that there is a building up of reserves from profits to protect against potential risks rather than depositing a share of the profits with the investment account holders.

References

Ahmad, K. (2003) An Islamic perspective, in: Dunning, J. (ed.), *Making Globalisation Good*, Oxford: Oxford University Press.

Ahmed, H., Mohieldin, M., Verbeek, J. and Aboulmagd, F.W. (2015) *On the sustainable development goals and the role of Islamic finance*. Retrieved from https://papers.ssrn.com/sol3/papers.cfm?abstract_id=2606839 (accessed 15 November 2017).

Al-Amine, A.M.M. (2013) Risk and derivatives in Islamic finance: A Shari'ah analysis. In: Hunt-Ahmed, K. (ed.) *Contemporary Islamic finance: Innovations, and best practices*. New Jersey: John Wiley & Sons, pp. 331–52.

Alrifai, T. (2015) *Islamic Finance and the New Financial System: An ethical approach to preventing future financial crises*, Singapore: John Wiley & Sons Singapore.

Askari, H., Iqbal, Z., Krichene, N. and Mirakhor, A. (2012) *Risk Sharing in Finance: The Islamic finance alternative*. Singapore: John Wiley & Sons Singapore.

Beekun, R.I. and Badawi, J.A. (2005) Balancing ethical responsibility among multiple organizational stakeholders: The Islamic perspective, *Journal of Business Ethics*, 60, pp. 131–145.

Chapra, M.U. (1992) *Islam and the Economic Challenge*. Leicester, UK: International Institute of Islamic Thought.

Daly, S. and Frikha, M. (2015) Islamic finance a support to development and economic growth: the principle of Zakat as an example, *Journal of Behavioral Economics, Finance, Entrepreneurship, Accounting and Transport*, 3(1), pp. 1–11.

Demuth, F. (2014) Parallel universes: What Islamic finance can learn from socially responsible investment, *Kyoto Bulletin of Islamic Area Studies*, 7, pp. 20–32.

El-Hawary, D., Grais, W. and Iqbal, Z. (2004) *Regulating Islamic Financial Institutions: The Nature of the Regulated*. World Bank Working Paper 3227, Washington, DC: World Bank.

Ernst, D., Akbiyik, D. and Srour, A. (2013) *Islamic Banking and Finance* Munich: UVK Verlagsgesellschaft mbH.

Greuning, V.H. and Iqbal, Z. (2008) *Risk Analysis for Islamic Banks*. Washington, DC: The International Bank for Remonstration and Development/The World Bank.

Hayat, U. and Malik, A. (2014) *Islamic Finance: Ethics, Concepts, Practice. A literature Review*. Charlottesville (USA): The CFA Institute Research Foundation.

Iqbal, Z. and Mirakhor, A. (2011) *An Introduction to Islamic Finance: Theory and practice*, 2nd edition. Singapore: John Wiley & Sons (Asia).

Iqbal, M. and Molyneux, P. (2005) *Thirty Years of Islamic Banking: History, performance, and prospects*. New York: Palgrave Macmillan.

IRTI (2015) *Islamic Social Finance Report 2015*. Retrieved from www.irti.org/English/News/Pages/IRTI-LAUNCHES-ISLAMIC-SOCIAL-FINANCE-REPORT-2015.aspx (accessed 20 December 2017).

Jusnaidi, N.F. (2015) Sustainability and role of theology: Islamic perspectives, *The Journal of Management and Science (AlQimah)*, 1(1), pp. 1–10.

Jusoff, K. and Alam, S. (2011) Environmental sustainability: What Islam propagates, *World Applied Sciences Journal*, 12 (special issue on Creating a Knowledge-based Society), pp. 46–53.

Kettell, B. (2011) *Introduction to Islamic Banking and Finance*. West Sussex, UK: John Wiley & Sons.

Kirchner, S. (2011) Faith, ethics and religious norms in a globalized environment: Freedom of religion as a challenge to the regulation of Islamic finance in Europe, *Baltic Journal of Law & Politics*, 4(1), pp. 52–82.

Lewis, K.M. (2009) In what ways does Islamic banking differ from the conventional finance, *The Journal of Islamic Economics, Banking and Finance*, pp. 9–24.

Malaysian Islamic Finance Centre MIFC (2016) *SRI and Green Sukuk – Challenges and Prospects*. Retrieved from www.sukuk.com/article/sri-and-green-sukuk-challenges-and-prospects-4762/ (accessed 20 December 2017).

Malkawi, H.B. (2014) Financial derivatives between Western legal tradition and Islamic finance: a comparative approach, *Journal of Banking Regulation*, 15(1), pp. 41–55.

Mohamad, S., Othman, J., Roslin, R. and Lehner, M.O. (2014) The use of Islamic hedging instruments as non-speculative risk management tools, *Venture Capital: An International Journal of Entrepreneurial Finance*, 16(3), pp. 207–26.

Nienhaus, V. (2015) Self-adjusting profit sharing ratios for Musharakah financing. In: El-Karanshawy *et al.* (eds) *Islamic Banking and Finance–Essays on corporate finance, efficiency and product development*, Doha: Bloomsbury Qatar Foundation, pp. 161–72.

Olaoye, I.K., Dabiri, M.A.F. and Kareem, R. (2013) Islamic banking in Nigeria: Challenges and prospects, *International Journal of Management Sciences*, 1(1), pp. 24–29.

Tavanti, M. (2013) Before microfinance: The social value of micro savings in Vincentian poverty reduction, *Journal of Business Ethics*, 112(4), pp. 697–706.

Ünal, M. (2013) The best of all worlds: Towards a more sustainable financial system. Retrieved from www.funds-at-work.com/en/intelligence/research-publications.html (accessed 27 September 2013).

Usman, A.S. and Tasmin, R. (2016) The relevance of Islamic micro-finance in achieving the Sustainable Development Goals, *International Journal of Latest Trends in Finance and Economic Sciences*, 6(2), pp. 1115–1125.

Williams, G. and Zinkin, J. (2010) Islam and CSR: A study of the compatibility between the tenets of Islam and the UN Global Compact, *Journal of Business Ethics*, 91(4), pp. 519–533.

8

FAITH-BASED ORGANIZATIONS AND CORPORATE SUSTAINABILITY

Joyce Stubblefield and Jan Jaap Bouma[1]

In many ways, sustainability can be a part of business strategies. Can we expect that the faith-based organizations (FBOs) will play a significant role in this internalization process? FBOs have many diverse interests in economic behaviour. One can argue that faith and Corporate Social Responsibility (CSR) can provide new paradigms for corporate sustainability. This thought opens a window of opportunities, as religions are often important factors in economic behaviour (Hui, 2008). FBOs – endowed with religious capital – can be explored as a capacity of societies to stimulate a more sustainable world. Also, the rise of CSR may facilitate FBOs in their efforts to promote sustainability. This chapter discusses drives by FBOs to innovate business models with a view to becoming more sustainable. To perform this assessment, a conceptual framework is presented that focuses on the intersection between religious capital and the implementation of sustainable business models.

Religious capital

Lynn White's 1967 article, The Historical Roots of our Ecological Crisis, brought diverse religious groups together to become involved in solving or seeking to solve the crises society has caused, even while some disagreed with White's laying the blame for the crises at the feet of religion as he stated: 'Hence we shall continue to have a worsening ecologic crisis until we reject the Christian axiom that nature has no reason for existence save to serve man. Since the roots of our trouble are so largely religious, the remedy must also be essentially religious, whether we call it that or not' (White, 1967). In the early 1970s, the World Council of Churches began to more fully and globally advocate for sustainable societies alongside the United States society's awakening to a rising pollution crisis (Earth Day 1970).

In 1995, Prince Philip created the Alliance of Religion and Conservation (ARC), 'to link secular worlds of conservation and ecology with faith worlds of major religions'; the ARC then 'asked faith leaders what they saw as the biggest problems for them and their tradition in terms of contemporary society' and they cited two things:

1   Mass communication, especially satellite TV with its particular brand of Western values,
2   The power of economic forces such as the World Bank.[2]

In 1997, a meeting organized by the ARC, which was held at the Archbishop of Canterbury's London Palace, discussed how 'alternative economic models arising from the faiths could help reduce poverty and environmental destruction'. That discussion resulted into various ARC/World Bank projects.[3]

Religious organizations have made efforts (plans, actions) and appeals (encyclicals, declarations, commitments) to promote environmental and economic sustainability. To accomplish that, these organizations must deal with complex issues that affect organizational finance, leadership roles, appropriate actions, shared networks as well as communication. Dialogue should generate and accelerate the work of the church on promoting a liveable, equitable and sustainable world.

The attributes of religious capital can be used to facilitate innovations in business models that reflect the values in accordance with sustainability; to be effective, these values should be shared by the firm's stakeholders. These stakeholders can be local, regional or even global, depending on the scale of operation and the nature of the issues at stake. Climate change is more global, whereas the scarcity of natural resources such as water and food may be more of a local or regional interest.

In 2009, the ARC, in partnership with the United Nations Development Programme, devised a strategic approach that resulted in 31 long-term commitments to environmental action by nine faiths worldwide. In this context, interfaith meetings may speed up commitments and actions. The Quakers provide examples of these commitments, when stating that 'Sustainability is a core spiritual and strategic priority for Quakers in Britain' and that 'Sustainability is an urgent matter for our Quaker Witness'. These statements derive from the plan to engage in interfaith dialogue on climate-induced migration.[4] Their seven-year plan can be accessed at: https://files.acrobat.com/a/preview/320c3835-d5de-4bd3-8b4d-f1a210515ba9.

In 2013, the ARC partnered with the Club of Rome on a radical programme, called Values Quest, a social, cultural and philosophical enquiry to uncover and challenge values underpinning contemporary European society, to motivate change. Pope Francis's 2015 encyclical called for moral leadership and dialogue on sustainability. In September 2015, at a meeting of the new United Nations Sustainable Development Goals (SDGs), faith leaders met and pledged to work to address the 'Bristol Commitments to help the world's poorest people'. These commitments included plans to develop micro-credit schemes for the poor, to increase access to education, to plant trees, to invest in clean energy and to engage in green pilgrimages.

The meetings cited thus by far were not the only meetings designed to address these issues. Different other religious groups organized other conferences both ahead (as in 2009) and after the 2015 Paris Climate Change Conference.

From the 1970s to the present, it has been noted that religious organizations' business models and social networks have played critical roles in managing and implementing sustainable actions of individuals and organizations, which can have complicated implications for balancing the four pillars of sustainability: people, planet, prosperity and time. They were challenged to expand the conversation and to engage within society about sustainability or sustainable development needs. The world community, businesses and religions have been wrestling with sustainability through the lens of concepts such as (1) corporate citizenship and social responsibility; (2) the development and implementation of standards; and (3) the evolution of strong grassroots organizations addressing issues of environmental justice and climate change legislation.

As stated by Martin Palmer, director of the ARC in 2009:

> [T]he great strength of the faiths is that they are the most sustainable of all human institutions, having outlived empires, dynasties and ideologies. This longevity means that they also take the time to change. The commitments were, in many cases, major long-term plans, which will take time to unfold and to go through the process of internal debate and discussion. Contrary to popular belief, few faiths are fully hierarchical; most have extensive systems for consultation.[5]

In the context of this noted strength, "religious investors have been called the third largest group of investors in the world" (Van Cranenburgh et al., 2014).

The Quakers and Methodists have shown responsibility in the context of responsible investments (RIs) and helped to pioneer modern forms of RI. In the literature on CSR, public interest in these religious groups and their societal roles is growing (Louche et al., 2012). The religious organizations' social concerns are 'about how to harmonize the values they preached and the realities of business activity with their traditions, such as the Jewish doctrine or the Catholic traditions'. Notable facts (Louche et al., 2012) are:

1   John Wesley, the founder of Methodism, emphasized that the use of money was the second most important subject of New Testament teachings.
2   Quakers have refused to invest in weapons and slavery.
3   Other religions have also engaged in RI (e.g., in Islamic banking).

Schueth (2003) also provided similar statements by John Wesley. In addition, he stated that 'It was likely that Methodist and Quaker immigrants brought the concept of social responsibility in investments to the new world. Since its beginning, over 200 years ago, the practice has been alive (Schueth, 2003). Deep religious roots can still be seen in the concept of 'sin stocks' as recognized by the majority of socially conscious investors in the USA (Schueth, 2003).

Sustainable decision making

In the context of sustainable business behaviour, business models are guiding the decisions of organizations, while sustainable business models focus on resolving social and environmental issues. Therefore, profit generation is unlikely to be their paramount concern (Dentchev *et al.*, 2016). Bocken *et al.* (2014) consider a sustainable business model as one that has incorporated the triple bottom-line approach (People, Planet, Profit) with implications for competitive advantage.

To compare business with religious organizations, the following points can be made (Louche *et al.*, 2012):

1   Religious organizations were the pioneers of RI.
2   Some of these have established corporations to facilitate business activities, whereas others operate through unincorporated associations.
3   They offer significant and substantial investments.
4   There are commonalities with the secular investors. According to Louche *et al.* (2012), religious organizations have networks with limited connections with the RI community; moreover, little is known about who creates policy or has investment power decisions.

Hui (2008) presented a value-driven framework that delivers value creation based on faith and CSR. According to Hui (2008), this model could help companies embed their perceived moral values in their strategic decision making. Also, the concept of intent is reflected here; an established strategic intent, shared by all levels of a firm, can cause people to perform in a way that unseats the best, but there is no guarantee. Similarly, in different religions, the most important moral laws cannot always help people to do the right things. Along the same line of reasoning, the models of corporate sustainability advocated by religious organizations do not guarantee that people act in accordance with the so-called journey towards a sustainable society. In other words, there is a difference between knowing and keeping the law as embedded in the rules of the game (Hui, 2008). Nevertheless, FBOs in various circumstances can be said to contribute to sustainable value creation.

According to Fry and Slocum (2008), for leaders there lies a great challenge in the development of new business models that embrace ethical values in leadership and the incorporation of sustainability into the core of management's operating portfolio. They stated that 'corporate performance is linked to strong ethical leadership' and provided a Model for Spiritual Leadership, which includes CSR. Here, spirituality can be with or without religion.

Moreover, Lozano (2012) introduced voluntary initiatives that embrace the four dimensions of sustainability: economic, environmental, social and temporal. His discussion included the merits of CSR, Environmental Management Systems, Sustainability Reporting and Corporate Citizenship, which could play a role in business models (Lozano, 2012). Louche *et al.* (2012) noted that the practice of RI is an indication of the presence of a substantial CSR documentation. Fry and Slocum (2008) provided the following definition of a business model: 'A business model is a description of the value a company offers to one or several sets of customers. It is the architecture of the firm and the network of partners/stakeholders'. These authors also called for social connection and a sense of calling as central themes to spiritual leadership processes (Fry and Slocum, 2008). In this sense, this business model definition is akin to the meaning of religious capital.

Indeed, there are elements in organizational theory that integrate the performance dimensions of sustainability (People, Planet, Profit) using the attributes of religious capital. Where religious leaders play an outspoken role in defining an organization's business model, that organization can be called a religious enterprise.

The religious enterprise

A religious enterprise may belong to any religion. In the financial sector, religious enterprise is primarily associated with Islamic banking, but this is not a complete representation of the state of affairs. Moreover, in a variety of economic sectors, the role of FBOs is recognized as a driver towards sustainability. Each FBO may have its capabilities to fulfil this facilitating role. Nevertheless, an FBO may also create or consolidate particular hindrances to this process. Therefore, there is no consensus on all performance dimensions of a sustainable enterprise.

Bitange Ndemo (2006) captured Weber's account that Protestantism supplied the moral energy and drive of the capitalistic entrepreneur; however, a Christian in a business or a Christian running a business does not guarantee the business is Christian (Dodd and Seaman, 1998). Bitange Ndemo (2006) considered social interventions by FBOs as being motivated by Mathew 25: 31–46; it is about a call to help others. La Barbera (1992) identified a study according to which spiritual missions and goals are often the primary motivation for both Jewish and Christian organizations to engage in commercial enterprise.

Argandoña (2012) contextualized a religious enterprise as a means for corporate social responsibility (CSR) by reviewing Pope Benedict XVI's Encyclical, Caritas in Veritate (2009), but evaluated that the encyclical's stance on CSR was only marginal. He went on to report that the encyclical proposed 'a profoundly new way of understanding business enterprise' (Benedict XVI, 2009). To attune to higher human and societal aims, the 'logic of gift', the 'principle of gratuitousness' and the 'division of labour' between the market, state and civil society come into the equation.

There is a need for a market that:

> permits the free operation, in conditions of equal opportunity, of enterprises in pursuit of different institutional ends [and this] requires that shape and structure be given to those types of economic initiative which, without rejecting profit, aim at a higher goal than the mere logic of the exchange of equivalents, of profit as an end in itself.
>
> Benedict XVI, 2009

Klein *et al.* (2017) posited that religious enterprises are universal and that values imbued by the religious or non-religious could operate in both religious and non-religious enterprises. Religious enterprises inspired by belief systems and a higher calling offer a frame for discourses about religious enterprising.

Terms and definitions

The literature provides various definitions of a religious enterprise: social entrepreneurship, social intervention by FBOs, FBO ventures and enterprises (Bitange Ndemo, 2006); religious social service (philanthropic and benevolent acts); social venture enterprise or faith-saturated social venture enterprise (Starling, 2010); and religious-motivated enterprises (RMEs) (Klein *et al.* 2017). Starling (2010) defined a social venture enterprise as 'a hybrid of business and non-profit best practices merged for meeting social needs. It is a new concept in the marketplace because it is a non-profit, earned-income business, and it is creating a fourth sector in the U.S. economy'. Klein *et al.* (2017) referenced La Barbera's study and defined an RME, for the purposes of their paper, as 'organizations providing goods and services to the marketplace that include objectives and values directly attributable to religious doctrine in their mission and the needs of their members'. La Barbera (1992) aligned a religious enterprise to the sacred/secular continuum model. She asserted that this model is identified by a plethora of historians, anthropologists, sociologists and psychologists (for example, Durkheim, 1912; Hammond, 1985), (as) relevant to contextualizing the sphere of this concept.

The interface between FBO and CSR

FBOs contribute to the religious capital of societies and facilitate the creation of values. The valuation of the underlying attributes may be different among organizations. Individual preferences can play an important role. Differences between communities, regions and countries do exist. FBOs may within their internal processes shape their approach to CSR and religious enterprise. Sustainable business models can help offer sustainable solutions and, if operated system-wide, these may have a notably positive impact on society. These connections become

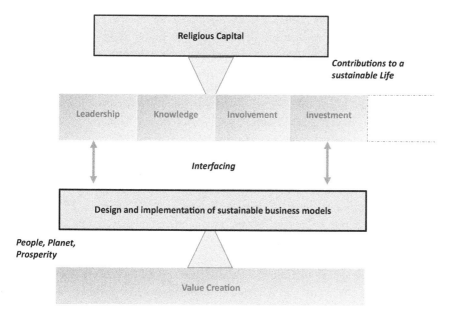

FIGURE 8.1 Interfacing concepts to address social norms and pressures of scale for sustainable solutions

visible by presenting variables placed on the intersection between religious capital and innovations in improved business models in Figure 8.1. These intersections provide stimuli to further innovate in the interface between FBOs and CSR.

Concluding remarks

If and how FBOs manifest themselves to facilitate sustainable companies is dependent on the specific context and changes over time. They can play the role of a community problem solver involving new attitudes towards sustainability and best practices in integrating environmental management standards and codes of conduct. As a result, companies may take on board CSR, ethical investments and extended producer responsibility while disclosure and reporting become complementary to their businesses.

It appears that the religious sector can combine values such as social peace, justice, equity, development and stewardship with **sustainable acumen** throughout its global and local networks. This is highly relevant considering that sustainable business is one of the most pressing intellectual and practical challenges facing world business, surpassing even global financial issues (Mort, 2010).

Religious organizations, serving billions of adherents, cannot ignore the sustainable business challenges faced by their members in addressing society's ills.

Religious leaders need to enhance stewardship based on their sustainable business acumen, leading to investments in the environment and in society at large.

Religious leaders are lobbying, marching, declaring and taking action to reduce negative environmental, societal and economic impacts and to attend to the needs of the most vulnerable in society. As new financial decisions are being considered across the marketplace to boost clean energy, improve energy efficiency and promote green building architecture, and protect natural habitats; the religious sector appears to be increasingly engaged in financial decisions that lead to ethical and sustainable outcomes for all.

Also, religious organizations appeared to have utilized the concept of *sustainable development* as a way to call for an equitable resource distribution that cares for poor societies and for a roadmap to address issues of climate injustice. FBOs may stimulate dialogues that shape sustainable actions among religious groups. A major example of this is the 2015 statement by Pope Francis (2015):

> I urgently appeal, then, for a new dialogue about how we are shaping the future of our planet. . . . We need a conversation which includes everyone, since the environmental challenge we are undergoing, and its human roots, concern and affect us all.

Sustainable leadership proved by FBOs may be regarded as *calls for action* within both secular and non-secular sectors. Religious organizations are facing increasing responsibilities as the world's human population continues to expand at the rate of more than 70,000,000 net increase per year, and as societies confront increasing food and water insecurities due to climate changes, political upheavals and poverty.

Some religious leaders are unsure about what to do for the benefit of sustainability while others ignore the subject. Because of that, in certain cases, sustainability-inspired church members may struggle to involve church leaders in sustainability-led networks outside the church's formal networks.

Religious capital can facilitate entrepreneurs who ask themselves how to design and implement sustainable business models. They can reflect on how leadership can shape sustainability in their organization. Also, faith groups can continue to play roles in RI – for instance, in the area of screening investment opportunities. Malloch (2008) pioneered the concept of 'spiritual capital' while Hui, Fry and Slocum Jr. captured spiritual leadership and value creation models. Encouraging sustainable business models by religious communities can significantly advance their understanding and application; such efforts may have an impact on society that goes beyond the current state of religious organizational actions.

Also, entrepreneurs may play a role in stimulating dialogues among different FBOs about innovative business models. For example, the Methodist and Quaker organizations work across religious traditions to collectively take sustainable action that includes business decisions and partnerships.

Notes

1   Acknowledgments. Our special thanks go to Professor Donald Huisingh (Tennessee) and Reverend Pat Watkins (Georgia) for their reflections and critical contributions to earlier versions of this chapter.
2   www.arcworld.org/about.asp?pageID=2#95
3   www.arcworld.org/about.asp?pageID=2
4   www.arcworld.org/downloads/Christian-Quakers.pdf
5   www.arcworld.org/downloads/One%20year%20after%20Windsor%20report%20 Nov%202010.pdf

References

Argandoña, A. (2012) Corporate social responsibility in the encyclical Caritas in Veritate. In: Schlag, M. and Mercado, J.A. (eds) *Free Markets and the Culture of Common Good*, Dordrecht, NL: Springer.
Benedict XVI (2009) *Caritas in Veritate*. Encyclical letter.
Bitange Ndemo, E. (2006) Assessing sustainability of faith-based enterprises in Kenya' *International Journal of Social Economics*, 33(5/6), pp. 446–62.
Bocken, N.M.P., Short, S.W., Rana, P. and Evans, S. (2014) A literature and practice review to develop sustainable business model archetypes, *Journal of Cleaner Production*, 65, pp. 42–56.
Cranenburgh, K. Van, Arenas, D., Goodman, J. and Louche, C.L. (2014) Religious organisations as investors: A Christian perspective on shareholder engagement, *Society and Business Review*, 9(2), pp. 195–213.
Dentchev, N., Baumgartner, R., Dieleman, H., Jóhannsdóttir, L., Jonker, J., Nyberg, T., Rauter, R., Rosano, M., Snihur, Y., Tang, X. and Van Hoof, B. (2016) Embracing the variety of sustainable business models: social entrepreneurship, corporate intrapreneurship, creativity, innovation, and other approaches to sustainability challenges, *Journal of Cleaner Production*, 113, 1–4.
Durkheim, E. (1912) *The Elementary Forms of the Religious Life: A Study in Religious Sociology*. A new translation from the French language by C. Cosman (2008), Oxford: Oxford University Press.
Pope Francis (2015) *Encyclical letter Laudato Si of the holy father Francis on care for our common home*. Section 63. Available at: http://w2.vatican.va/content/dam/francesco/pdf/ encyclicals/documents/papa-francesco_20150524_enciclica-laudato-si_en.pdf
Fry, L.W. and Slocum, J.W. (2008) Maximizing the triple bottom line through spiritual leadership, *Organizational Dynamics*, 37(1), pp. 86–97.
Hammond, P.E. (1985) *The Sacred in a Secular Age: Toward Revision in the Scientific Study of Religion*. Berkeley, CA: University of California Press.
Hinings, C.R. and Raynard, M. (2014) Organizational form, structure, and religious organizations', *Research in the Sociology of Organizations*, 41, pp. 159–186
Hui, L.T. (2008) Combining faith and CSR: a paradigm of corporate sustainability, *International Journal of Social Economics*, 35(6), pp. 449–65.
Klein, T.A., Laczniak, G.R. and Santos, N.J.C. (2017) Religion-motivated enterprises in the marketplace: A macromarketing inquiry, *Journal of Macromarketing*, 37, pp. 102–114.
La Barbera, P.A. (1992) Enterprise in religious-based organizations, *Nonprofit and Voluntary Sector Quarterly*, 21(1), pp. 51–67.
Louche, C., Arenas, D. and Cranenburgh, K.C. van (2012) From preaching to investing: Attitudes of religious organisations towards responsible investment, *Journal of Business Ethics*, 110(3), pp. 301–20.

Lozano, R. (2008) Envisioning sustainability three-dimensionally, *Journal of Cleaner Production*, 16, pp. 1838–46.

Lozano, R. (2012) Towards better embedding sustainability into companies' systems: An analysis of voluntary corporate initiatives, *Journal of Cleaner Production*, 25, pp. 14–26.

Malloch, T.R. (2008) *Spiritual Enterprise: Doing virtuous business*. New York: Encounter Books.

Mort, G.S. (2010) Sustainable business, *Journal of World Business*, 45(4), pp. 323–5.

Muehlhausen, J. (2013) *Business Models for Dummies*, Hoboken, NJ: J. Wiley & Sons.

Schueth, S. (2003) Socially responsible investing in the United States, *Journal of Business Ethics*, 43, 189.

Starling, V.T. (2010) *Faith-based Social Venture Enterprise: A model for financial sustainability*. Minneapolis, MN: Walden University.

White Jr, L. (1967) The historical roots of our ecologic crisis, *Science*, 155.

9

THE GREENING OF CHURCHES

A maturity framework based on policies and measures

Ankeneel A. Breuning

Introduction

In many congregations, people are working on a more sustainable church, both in terms of how the community operates and in terms of the energy use of the building. The primary motive behind this movement is religious in accordance with the Christian teaching (van der Spek and Olbertijn, 2011; Pope Francis, 2015). An additional motive is a reduction in energy costs that helps balance budgets.

In The Netherlands, for many years, the Green Churches programme (*Groene Kerken*) has been a steady promotor of sustainability and how religious communities could embrace it. This programme gives support to local churches (including a number of mosques) in their quest for a sustainable organization; it provides an online platform with a wide variety of information on what kind of green measures are possible, organizes conferences to exchange information and gives courage to those committed to the cause (see www.groenekerken.nl).

Even though there is support, becoming a sustainable church remains difficult. While there is a desire to act, taking concrete action can be challenging because of a lack of (technical) knowledge and limited economic resources. Moreover, there is no standardized approach to creating a sustainable church beyond the common understanding that sustainability is about creating a balance between economic, ecological and social factors. Therefore, this chapter takes up the task of developing and presenting a method that could at least be a major step towards such a standardized approach. It should define what a sustainable church is and provide a way to gauge a church building's current performance and assess which potential improvements are available. This chapter introduces a scan to deal with various issues. This scan is not just a matter of inspecting a church building from a technical point of view; it also takes into account the policy and management aspects of maintaining and running it.

Objectives and scope

A building's characteristics of a church depend on factors such as the kind of denomination, size, shape, building period and style (Roeterdink *et al.*, 2008). Church buildings vary considerably, both in appearance and in the way they are used. Most church buildings are focused on Sundays as a day of worship, but this is not a uniform pattern in all cases (Ketel, 2012; Ketel and Scheele, 2014). Traditionally, churchgoers make use of pews or typical chairs, but nowadays this is not common procedure in every congregation. When considering what can be done to further sustainability, such characteristics need to be taken into account.

Methodology

Background

In the world of business, the Triple Bottom Line (TBL) is a model widely used to make sustainability practical (see Elkington, 1999). A similar model sees sustainability as having three pillars, that is, economic vitality, environmental integrity and social equity (Kastenhofer and Rammel, 2005).

More recently, in certain cases, the TBL, also referred to as People, Planet, Profit (the 3P's), replaces Profit by Prosperity. According to the TBL, sustainability can only be achieved

> when there is a reconciliation between (1) economic development; (2) meeting, growing and changing human needs and aspirations on an equitable basis; and (3) conserving limited natural resources and the capacity of the environment to absorb the stresses that are the consequence of human activities.
>
> Hay and Mimura, 2006

However, this line of reasoning can only become relevant to church buildings if there is a clear connection between the TBL and Sustainable Building (SB).

SB requires understanding a building's impacts on the environment and tries to amplify the positive and mitigate the negative impacts over a building's life cycle. Furthermore, sustainability in terms of the TBL envisages a continuous integration of a building's economic, social and environmental performance during its whole life time (John *et al.*, 2005). A building's integrated quality can be expressed by adding a 'technical' pillar to the TBL framework (Hill and Bowen, 1997; Akadiri *et al.*, 2012).

In SB, energy demand during the user phase is most significant. A commonly accepted strategy for energy management in SB is using the Trias Energetica as an assessment tool (Lysen, 1996). The Trias Energetica consists of the following three steps: Prevention, Renewable and Efficiency (Table 9.1). The assessment of measures must be in conformity with the TBL.

TABLE 9.1 Trias Energetica

No.	Name	Description	
1	Prevention	Reduce energy demand by preventing the use of energy	
2	Renewable	Use sustainable, renewable sources	
3	Efficiency	Meet the remaining energy demand (requiring fossil fuels) as efficiently as possible	

Lysen, 1996; Entrop and Brouwers, 2010

Most assessment tools do not apply to a sustainable church building, due either to the absence of economic factors in the the analysis (Ding, 2008) or because the tools do not fit with the specific building type. Tools that do match with the three TBL pillars include the DGNB certification system, SBTool and EcoProp (Larsson, 1998; Ding, 2008; Mateus and Bragança, 2011; Andrade and Bragança, 2016), as well as a promising pilot upgrading of LEED to the TBL ('New LEED Pilot Credit for TBL Analysis', n.d.). Even though a tool is not (fully) applicable, various components could provide new insights. There are three major components: (1) the Structure, (2) the Scoring and (3) the Output (Cole, 1999).

The Structure builds on the adopted criteria and is measured or predicted through the use of sustainability indicators (Hammond et al., 1995). There is, as yet, no unambiguous technique to select sustainability indicators. Standardized sets are available (Table 9.2); however, often, these sets are considered incomplete (Fernández-Sánchez and Rodríguez-López, 2010); whether they are suitable depends on a project's stakeholders (Alwaer and Clements-Croome, 2010). Scoring indicators remain a controversial aspect; in particular, the absence of an agreed theoretical basis for deriving suitable weighing factors is a concern (Cole, 1998). Four types of comparison can help to observe a reasonable level of objectivity:

1 Comparing a specific performance criterion relative to a declared benchmark.
2 Comparing performance scores of one criterion to those of other criteria within the same building.
3 Comparing the performance of two buildings based on the same performance criterion.
4 Comparing the overall performance profile to other buildings' performance profiles (Cole, 1999; Haapio and Viitaniemi, 2008).

The outputs are often expressed as a single composite index, through numerical integration (Lindholm et al., 2007; Bell and Morse, 2013), which indicates the sustainability of the building. Normalization, possible standardization and aggregation are probable steps (Olsthoorn et al., 2001). In construction, there is an increasing use of multi-criteria analysis methods (MCAMs) (Ding, 2008; Jato-Espino et al., 2014).

TABLE 9.2 Sample of standards in the construction industry (updated version)

Standard	Standard title	published
ISO 21929–1:2011	Sustainability in building construction – sustainable indicators, Part 1: Framework for development of indicators for buildings.	Nov 11
ISO 21930:2007	Sustainability in building construction – environmental declaration of building products.	Oct 07
ISO 21930	Sustainability in buildings and civil engineering works – core rules for environmental declaration of construction products and services used in any type of construction works.	Draft
ISO 21932	Sustainability in building construction – terminology.	May 08
ISO 21932:2013	Sustainability in buildings and civil engineering works – a review of terminology.	Nov 13
ISO 15392:2008	Sustainability in building construction – general principles.	May 08
CEN EN 15643–1	Sustainability of construction works – integrated assessment of building performance. Part 1: a general framework.	Dec 10
CEN EN 15643–2	Sustainability of construction works – integrated assessment of building performance. Part 2: framework for the assessment of environmental performance.	May 11
CEN EN 15643–3	Sustainability of construction works – integrated assessment of building performance. Part 3: framework for the assessment of social performance.	Apr 12
CEN EN 15643–4	Sustainability of construction works – integrated assessment of building performance. Part 4: framework for the assessment of economic performance.	Apr 12
NEN 7120	Terms, definition, and methods for the calculation of energy performance and related indicators of a building or part of a building. Usable for dwellings, residential and utility buildings, new as well as existing.	Oct 12

Fernández-Sánchez and Rodríguez-López, 2010

Presenting the outcome as a single index can be either undesirable or impossible. However, here, the application of maturity levels is an alternative, as Baumgartner and Ebner (2010) formulated for corporate sustainability. Maturity levels provide a quick insight into the current measures and potential measures. Linking maturity levels of sustainability to SB can be partly based on the existing sustainability strategies (involving intentions and running policies), but should also express the actual level of (financial) investment in sustainability measures and the level of integrated sustainability.

Research design

The main research was: 'How can the sustainability performance of churches in religious use be assessed and be improved?' The first step of the project was to develop a framework that could describe the sustainability of a church building, considering its characteristics and the view on sustainability of the community concerned. The second step was testing the framework by means of case studies, leading to an improved sustainability assessment framework. Finally, the third part presents an application of the assessment framework in an integrated approach including the current solution approaches.

Figure 9.1 presents a visualization of the research design.

[A] is the development of the sustainability assessment framework for churches based on the available literature. [B] is the practical part of the study, that is, the framework that was tested by applying it to three congregations and their church buildings in the city of Apeldoorn, resulting in an improved sustainability assessment framework [C]. [D] presents a(n) (additional) literature study on improving the energy efficiency of church buildings. [E] is the final applicable model containing the sustainability framework and various strategies and models for churches. [E] can be the basis for an integrated approach towards a sustainable church.

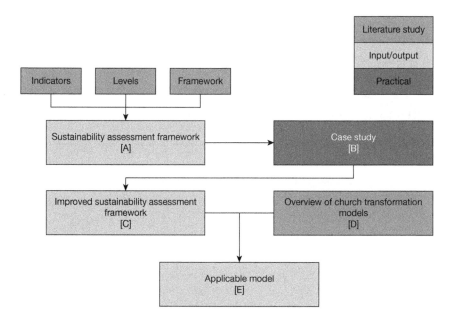

FIGURE 9.1 Visualization of the research design

Results

Framework

For the development of the basic framework to assess and improve the sustainability performance of a church, individual measures were studied as well as how these should come together in an improvement project. Here, the Trias Energetica appeared to be helpful; it distinguishes four categories:

1 Energy Management: human factors that influence the energy consumption.
2 Energy Demand: heating and cooling demand stemming from the characteristics of the building envelope.
3 Energy Efficiency: energy consumption due to equipment or energy-using systems.
4 Renewable Energy: the procurement of sustainable energy or generation onsite.

Each category covers a wide variety of measures; a detailed list of measures is available (Breuning, 2017). The criteria often used for church buildings are the investment costs, the energy costs, the impact on the monumental value, the preservation factor and the building's thermal comfort (Schellen, 2002; Camuffo, 2006; Ketel and Scheele, 2014). Striking a balance between conservation/preservation of the building (and its interior) and thermal comfort is common procedure. However, this kind of balancing neglects the ecological dimension, in particular, the potential reduction of greenhouse gas (GHG) emissions. Table 9.3 provides an overview of the indicators used for each of the three dimensions (economic, ecological and social).

TABLE 9.3 Overview of indicators for the three dimensions, including concise descriptions

Dimension	Indicator	Description
Economic	Investment Cost	An estimation of the investment cost.
	Energy Cost Reduction	An estimation of the potential savings on energy cost.
Ecological	GHG Emission Savings	The potential reduction of emissions.
Social	Monumental Value Impact	The impact on the monumental value of the building.
	Preservation	The effects of the indoor climate and the preservation.
	Thermal Comfort	An estimation of the effects on the thermal comfort.

TABLE 9.4 Maturity levels of sustainable performance of church buildings

Level	Description
0: Starting Point	No awareness
1: Initiated	The religious community is becoming aware and gains insight into the current situation.
2: Managed	The formulation of goals, objectives and strategies.
3: Defined	The implementation of measures without a significant investment.
4: Institutionalized	Implementation of measures that require investment.
5: Optimized	Improved system; constant reviewing of actions.

Often, the (ultimate) effect of a measure depends on the combined effects of different measures that are taken simultaneously. Unfortunately, the knowledge on how individual measures influence each other is limited; therefore, it is difficult, if not impossible, to single out the effects of one particular measure. It was, however, possible to circumvent such complications by defining several maturity levels to express the sustainability performance of a church building. There were six levels distinghuished, starting from level 0, where there is no awareness of the sustainability performance of the building, to the other end of the range, level 5, as the highest level. At the highest level, the church community has adopted an integrated approach to sustainability; that means, sustainability is part of the congregation's decision-making processes, leading to an optimal performance of the church building. Table 9.4 presents the various maturity levels (for more details, see Annex A).

A church community with its own church building that applies the model produces scores for each maturity level. However, maturity levels 2–5 can be taken into account only if the previous lower level has obtained at least 50% of maximum credit points. A list of 55 YES or NO statements determines the scores for the various sustainability levels: a statement that is supported by a YES adds a credit point.

Case study

As mentioned before, the case study took place in Apeldoorn. Three church buildings were included.

The first church building, the Assumption of the Virgin Mary Church (in Dutch: *Onze Lieve Vrouwe Ten Hemelopneming Kerk*, OLV), is a recognized national monument located in the centre of Apeldoorn. It is the only remaining Roman Catholic (RK) Church in Apeldoorn. Its construction started in 1895, the choir and transept in 1901; due to a lack of funds, the western tower remained unconstructed ('De Emmaüsparochie [The Emmaus Parish]', n.d.). The building's heating system was recently modernized.

The second church is the Open Hofkerk (OHK), a mainstream Protestant church (PKN) in the residential area of Kerschoten; this church building enjoys local monumental protection as part of Kerschoten, which has that status in its entirety. Over the past few years, the religious community of the OHK has experienced difficulties because of a diminishing and ageing congregation. It then developed an elaborate sustainability plan, with cost savings in mind, but the plan was not implemented due to a lack of funds. The church building is likely to be deconsecrated in the near future.

The third church is the Jachtlaankerk (JLK), which is a mainstream PKN in the upmarket area of De Sprengen in Apeldoorn West. The JLK was built in 1952 and officially opened in 1953. The year 2000 saw the completion of an extension, and more recently (2017) there was an analysis of the building's energy use, followed by a partial renovation that improved its sustainability performance. In contrast to the other two cases, the JLK's spaces are utilized for a variety of activities throughout the week, except on Saturdays.

Table 9.5 presents the outcome of the application of the framework, both for the original sustainability assessment framework and the improved version. The improvements in the framework stem from the cases and a discussion on their results by the Apeldoorn's Consultation Board of Churches' sustainability committee (which commissioned the research), The OLV reached level 1 only but also scored some credit points at the other levels, especially at level 4 (thanks to a review of the heating system). The OHK received high scores at all levels except level 5. This scoring showed that the framework attached too much value (weight) to projects that were planned but not (yet) implemented. This weakness was corrected in the revised framework. The JLK received very high scores for all levels of maturity (for further details, see Breuning, 2017).

An analysis of the content and checking of user validity formed the basis for the improved framework. The most significant change was less weight attached to intentions and more weight to the actual implementation of measures (see for the final list of statements Annex B). Table 9.5 shows the results of the improved assessment framework for the same cases.

TABLE 9.5 Results per maturity level from application of the framework

	OLV		OHK		JLK	
	Original	Improved	Original	Improved	Original	Improved
Level 0	x	**x**	x	x	x	x
Level 1	**71%**	80%	71%	60%	100%	100%
Level 2	22%	14%	100%	**100%**	100%	100%
Level 3	26%	27%	74%	73%	96%	100%
Level 4	56%	57%	**78%**	71%	100%	100%
Level 5	14%	0%	43%	33%	**100%**	**100%**

Application of the sustainability assessment framework

Religious communities in The Netherlands cannot escape the consequences of secularization: shrinking congregations and relatively high exploitation costs (Jelsma, 2012) result in seven possible strategies for retaining the building or, if this is impossible, its deconsecration (Table 9.6).

While Table 9.6 presents most of the available approaches, alternative publications provide other options especially in the area of alternative housing. Figure 9.2 gives a breakdown that includes these approaches. Retaining the building requires generating additional income, either by finding ways of reviving and increasing the congregation or by generating income from additional users who share the overheads. Closing down the building is likely to motivate a search for an alternative place of worship, which could be a matter of obtaining different accommodation or joining another congregation. The empty building may draw the attention of other organizations or communities, either religious or secular, interested in buying it.

The approaches as reflected in Table 9.6 do not consider sustainability, at least not as an explicit factor. This chapter advocates taking sustainability on board, even

TABLE 9.6 Transformation strategies

No.	Category	Description
I	New identity	A remake of the *traditional identity* of the religious community. A modern approach towards the identity and traditions to reach a *new (target) audience*.
II	Multifunctional use	*Multifunctional use of the church building without remodelling.* The building is next to the religious function used for a variety of cultural and social activities. This is *multifunctional use in time*.
III	Remodelling	*Remodelling of the church building for religious use.* Remodelling can modernize a layout that does not correspond with current requirements. There is often a combination with multifunctional use.
IV	Partial repurposing	*Remodelling of the church building maintaining a smaller religious space.* The new smaller place of worship allows for multifunctional use of the rest of the building for a wider variety of functions. This is an example of *multifunctional or shared use in space*.
V	Reusage	*The religious community vacates the building, but the function remains.* Another religious community uses the church building as their place of worship.
VI	Repurposing	*The religious community vacates the building and the function changes.* The new user transforms the church building for their own purposes.
VII	New building	A religious community decides to *construct a new building*.

Jelsma, 2012

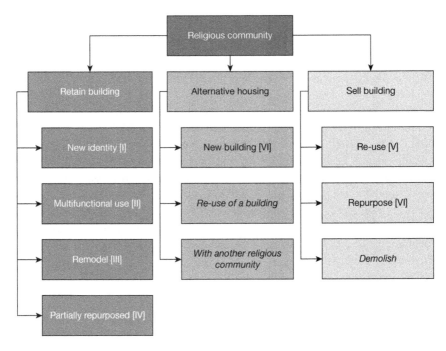

FIGURE 9.2 Adapted breakdown structure of the transformation categories

Jelsma, 2012; Stadig, 2016

though this is not without difficulties. To create sustainability-led investments there must be sufficient resources, in particular, time, expertise and funds. Especially, if a religious community struggles to make ends meet, a focus on sustainability may involve notable risks, especially, short-term financial risks. Studying sustainability-led projects involving churches that were completed elsewhere could be very helpful in recognizing the success and failure factors of such projects.

The proposed integrated transformation is problem-based (Figure 9.3). The structure starts with an analysis of the current situation (1), especially regarding the size of the church community and its financial situation. The left side of the structure presents the sustainability measures, which can be either preventive (2) or aimed at reducing energy costs (3.1). Both start by assessing the current sustainability situation (5), after which measures at different levels (5.1–5.4) can be taken, depending on the goal. Improving the occupancy efficiency (3.2) is another method used to increase the net income flow; two different types can be identified – multifunctional use (6.1) and partial repurposing (6.2), often implying less space for the congregation in order to facilitate a multifunctional income-raising programme. Finally, there is the option of moving into another building whose running costs are lower (3.3); simultaneously, options for disposing of the current building need to be explored (6) to define the financial scope for a new building. In regard to sale of the building, there are several options such as finding another

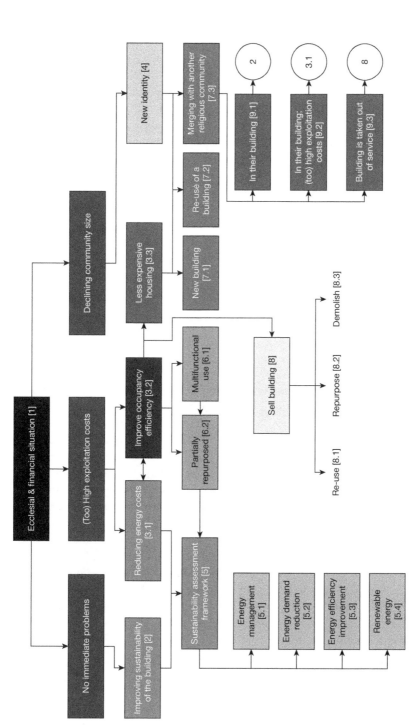

FIGURE 9.3 Proposed integrated approach for the transformation of church buildings

religious community interested in taking it over (8.1), selling it to another party who wishes to use it for a non-religious purpose (8.2) and, finally, demolishing the building and selling the site (8.3). How appropriate these options are depends on the various restrictions that new users have to take into account. By considering the three steps (3.1–3.3) together, combinations of options are presented. Changing a congregation's traditional identity (4) is hard to do but is possible in some cases. What happens quite often is a merger of two communities or parishes (5). This may be the only option that will make a real difference in terms of running costs. Here, it is important to look critically at both buildings and make a well-founded decision on what should be the best option for both communities (9.1–9.3).

An integrated approach as presented above has the advantage of selecting a solution that is geared to a specific community and its particular situation. Further research should look into the validity of the framework, especially as to the relationship between policies and measures taken. Moreover, it should be further explored whether the framework's measurements should be refined to justify firm conclusions and control risks. In any case, the model can be used to open a debate within a congregation on sustainability and what can be done to implement it.

Conclusion

Energy use is closely related to the sustainability performance of a church building, making it the focus of the framework. The energy use discussed in this chapter is based on absolute energy use, without considering the source. There are four categories (headings) of measures: (1) Energy Management; (2) Energy Demand; (3) Energy Efficiency; and (4) Renewable Energy. Six criteria, taken from the literature, define the TBL performance of a measure that can be taken: (1) Investment Costs; (2) Energy Cost Reduction; (3) GHG Emission Savings; (4) Monumental Value Impact; (5) Preservation; and (6) Thermal Comfort. As explained before, the interlinkages between measures prevent separate scores of an individual measure. This problem has been circumvented by using maturity levels. The six maturity levels describe the awareness of sustainability within the community, formulation of plans, the progress of implementation, and express the degree of integration of sustainability. The maturity levels are: (0) Starting Point; (1) Initiated; (2) Managed; (3) Defined; (4) Institutionalized; and (5) Optimized. An initial list of 55 statements records the actions taken by a religious community and determines the level of maturity (see Annex B). The results provide insight into the status of the building, and what further steps are considered. The developed framework is somewhat unique in bringing together policy making and technical performance as factors determining a church building's eventual sustainability performance. The case studies, however, illustrated that too much weight was given to the policy steps prior to the taking of concrete measures. This imbalance was corrected in the revised framework by reducing and amending the list of statements contained. Further empirical research on the framework's wider applicability is advisable in order to make the framework a resilient tool that can deal with a variety of situations while striking a balance between sound policies and actual measures taken. However, the framework as it is at present can already serve the purpose of initiating essential discussions on what it means to be a church community that takes sustainability seriously.

References

Akadiri, P.O., Chinyio, E.A. and Olomolaiye, P.O. (2012) Design of a sustainable building: A conceptual framework for sustainability in the building sector, *Buildings*, 2(2), pp. 126–152.

Alwaer, H. and Clements-Croome, D.J. (2010) Key performance indicators (KPIs) and priority setting in using the multi-attribute approach for assessing sustainable intelligent buildings, *Building Environment*, 45(4), pp. 799–807.

Andrade, J. and Bragança, L. (2016) Sustainability assessing of dwellings – a comparison of methodologies, *Civil Engineering and Environmental Systems*, 33(2), pp. 125–146.

Baumgartner, R.J. and Ebner, D. (2010) Corporate sustainability strategies, *Sustainable Development*, 18(2), pp. 76–89.

Bell, S. and Morse, S. (2013) *Measuring Sustainability: Learning from doing*. New York: Routledge.

Breuning, A.A. (2017) Towards a Sustainable Church, an exploratory research into the sustainability performance of church buildings in Apeldoorn. Master Thesis. Enschede, NL: University of Twente.

Camuffo, D. (2006) *Church Heating and the Preservation of the Cultural Heritage. Guide to the Analysis of the Pros and Cons of Various Heating Systems*. Milan: Mondadori Electa S.p.A.

Cole, R.J. (1998) Emerging trends in building environmental assessment methods, *Building Research & Information*, 26(1), pp. 3–16.

Cole, R.J. (1999) Building environmental assessment methods: clarifying intentions, *Building Research Information*, 27, pp. 230–246.

De Emmaüsparochie (The Emmaus Parish)' (n.d) Available at: www.emmaus-apeldoorn.nl/ (accessed 17 March 2017).

Ding, G.K.C. (2008) Sustainable construction – the role of environment assessment tools, *Journal of Environmental Management*, 86(3), pp. 451–464.

Elkington, J. (1999) *Cannibals with Forks; The Triple Bottom Line of 21st Century Business*, 2nd edition. Oxford: Capstone Publishing.

Entrop, A.G. and Brouwers, H.J.H. (2010) Assessing the Ssustainability of buildings using a framework of triad approaches, *Journal of Building Appraisal*, 5(4), pp. 293–310.

Fernández-Sánchez, G. and Rodríguez-López, F. (2010) A methodology to identify sustainability indicators in construction project management', *Ecological Indicators*, 10(6), pp. 1193–1201.

Groenekerken.nl – Kerken kiezen voor duurzaamheid (n.d). Available at: www.groene kerken.nl/ (accessed 3 November 2016).

Haapio, A. and Viitaniemi, P. (2008) A critical review of building environmental assessment tools, *Environmental Impact Assessment Review*, 28(7), pp. 469–482.

Hammond, A., Adriaanse, A., Rodenburg, E., Bryant D. and Woodward, R. (1995) *Environmental Indicators: A Systematic Approach to Measuring and Reporting on Environmental Policy Performance in the Context of Sustainable Development*. Washington, DC: World Resources Institute.

Hay, J. and Mimura, N. (2006) Supporting client change vulnerability and adaptation assessments in the Asia-Pacific region: An example of sustainability science, *Sustainability Science*, 1(1), pp. 23–35.

Hill, R.C. and Bowen, P.A. (1997) Sustainable construction: principles and a framework for attainment, *Construction Management and Economics*, 15(3), pp. 223–239.

Jato-Espino, D., Castillo-Lopez, E., Rodriguez-Hernandez, J. and Canteras-Jordana, J.C. (2014) A review of application of multi-criteria making methods in construction, *Automation in Construction*, 45(1), pp. 151–162.

Jelsma, T. (2012) Van traditioneel gebruik tot herbestemming – Transformatie als kans voor kerk en samenleving. In: *Meer Dan Hout En Steen – Handboek Voor Sluiting En Herbestemming van Kerkgebouwen*. Tweede druk. Zoetermeer: Uitgeverij Meinema.

John, G., Clements-Croome, D. and Jeronimidis, G. (2005) Sustainable building solutions: a review of lessons from the natural world, *Building Environment*, 40(3), pp. 319–328.

Kastenhofer, K. and Rammel, C. (2005) Obstacles to and potentials of the societal implementation of sustainable development: A comparative analysis of two case studies, *Sustainability: Science, Practice and Policy*, 1(2), pp. 5–13.

Ketel, D. (2012) *Kerkverwarming in Edese kerken; voor een beter binnenklimaat, minder CO2-uistoot en lagere energiekosten*. Ede: Ede's kerkelijk platform voor Duurzaamheid en Energiebesparing (EkDE).

Ketel, D. and Scheele, R. (2014). *Duurzame kerkverwarming* [Sustainable Church Heating]. Benekom: ZWO-taalgroep Kerk & Milieu/Samenleving.

Larsson, N. (1998) Research information: Green Building Challenge '98, *Building Research and Information*, 26(2), pp. 118–121.

Lindholm, O., Greatorex, J.M. and Paruch, A.M. (2007). Comparison of methods for calculation of sustainability indices for alternative sewerage systems – Theoretical and practical considerations', *Ecological Indicators*, 7(1), pp. 71–78.

Lysen, E. (1996) The Trias Energica: Solar Energy Strategies for Developing Countries. In: Goetzberger, A. and. Luther, J. (eds) *Proceedings of the EUROSUN Conference*, 16–19 September, Freiburg, Germany: DGS. Freiburg: Sonnenenergie Verlags-GmbH.

Mateus, R. and Bragança, L. (2011) Sustainability assessment and rating of buildings: Developing the methodology SBTool PTeH, *Building Environment*, 46(10), pp. 1962–1971.

New LEED Pilot Credit for TBL Analysis (n.d). Impact Infrastructure.

Olsthoorn, X., Tyteca, D., Wehrmeyer, W. and Wagner, M. (2001) Environmental indicators for business: A review of the literature and standardisation methods, *Journal of Cleaner Production*, 9(5), pp. 453–463.

Pope Francis (2015) *Laudato si*, encyclical letter on care for our common home.

Roeterdink, N., de Jong, J., Tilleman, L., Wierenga, R. and van Geel, P. (2008). *Onderzoek herbestemming kerken en kerklocaties; een inventarisatie vanaf 1970*. Bisdom Haarlem, Bisdom Rotterdam en Projectbureau Belvedere, Haarlem.

Schellen, H.L. (2002). *Heating Monumental Churches: Indoor climate and preservation of cultural heritage*. Eindhoven: Technische Universiteit Eindhoven.

Stadig, D. (2016) *Help, onze kerk loopt leeg! Een kleine handleiding voor kerkbestuurders*. Rijksdienst voor het Cultureel Erfgoed.

van der Spek, H. and Olbertijn, G. (2011) *Duurzaam ondernemen; Een praktisch werkboek* [Sustainable Enterprise; A Practical Workbook]. Putten: CBMC Nederland.

What is green building? U.S. Green Building Council (n.d). Available at: www.usgbc.org/articles/what-green-building (accessed 17 November 2016).

Young, J.W.S. (1997) A framework for the ultimate environmental index – Putting atmospheric change into context with sustainability, *Environmental Monitoring and Assessment*, 46(1–2), pp. 135–149.

Annex A: Defining the different maturity levels

Level 1: Initiated

The religious community is starting to raise awareness of the sustainability performance of their church building. There is active awareness when it comes to considering small measures, in particular. Also, the church is to start up monitoring and registering energy use. The measures at this initial level are needed to be able to move to higher levels; however small the effects on the actual energy usage, this level forms the foundation for a structured approach to the following levels. Even if the community had previously gone beyond this point of planning, it is advisable to return to the start. Returning to this point and formulating the status quo allows for evaluation of the measures, already implemented and/or planned.

Level 2: Managed

At this level, it is important that plans be defined. The religious community decides on a goal and defines the corresponding objectives and strategies. Forward-moving targets in the course of time will keep the community involved and aware of the project. It also allows for adjustment of plans, depending on the results of previously implemented measures. Equally important is the appointment of a person responsible for keeping track of the progress made and for signalling when things are going wrong. The second level plans initiate the steps to further levels and require continuous monitoring of the results of the implemented measures. To move to the next stage, goals and objectives must be clear; the strategies, however, can be less detailed, to prevent unnecessary work due to subsequent adjustments.

Level 3: Defined

The previous levels are mostly a matter of creating awareness of the importance of behavioural change and awakening a sense of urgency, suggesting an optimal pace of implementation in the near future. At this level, the implementation of measures gets off the ground. These measures do not need a significant investment level *per se*. Optimizing the control and operation of the system, as well as making small improvements to the building and its use, can be relevant actions. The actions will give extra insight into the characteristics of the building, which can serve to prepare for the subsequent implementation of greater measures. At this stage, it is also expedient to determine how sustainable energy can be procured. In any case, at this level, it is most important that concrete measures are taken. Therefore, this level sees a combination of preparatory activities and the actual taking of measures.

Level 4: Institutionalized

At this level, investments further improve the sustainability performance of the church building. Examples include improving the thermal conductivity of the building envelope or generating energy on-site. Attention could be paid to the kind of heating system that would be optimal, keeping in mind the building and occupancy characteristics and the state of the current installations. Plans should consider the interlinkages between measures and what that means for the eventual performance of the entire set of measures. Here, eventually, the end of the project will be approaching; there are detailed plans and significant implementations.

Level 5: Optimized

At this level, sustainability is part of every activity. Continued monitoring is essential to ensure that the system's performance is optimal. Possibly, there should be additional measures to reach the final goal. Maintenance will be necessary as well as a continued maintenance the project in order to prevent any kind of backsliding. Level 5 is the stage where the goals and objectives are reached. Here, the community should keep up to date with the new green technologies that may arise. It does not become stagnant but keeps progressing through optimization and renewal of the system.

Annex B: The questionnaire's yes/no statements

1 There is at least a monthly registration of energy use.
2 There is recurring attention given to sustainability within the religious community.
3 There is attention to sustainability within the church building.
4 There is enthusiasm created for the sustainability project within the religious community.
5 There is a policy for the use of the building.
6 There is a vision for the future and improvement of the sustainability of the building.
7 The objectives are specific, measurable, acceptable, realistic and time-bound.
8 There are short-term goals (< 1 year).
9 There are long-term goals (> 1 year).
10 There is a plan of action to attain the objectives.
11 There are scheduled evaluation moments to measure progress.
12 Appointment of an energy manager to supervise progress of the plans.
13 There is an inventory containing the possible improvements to the heating system.
14 The settings of the energy system are optimized to improve efficiency.
15 There is an exploratory study of the possibility of zoning the system.
16 Implemented zoning allows for independently controlled zones.

17 There is a study of the temperature profile of the church (hall), with measurements at a minimal interval of five minutes for a period of at least three weeks.
18 The occupancy of the building is known.
19 There is an optimized base temperature, depending on the measurement results and the use of the building.
20 There is a scheduling policy of planning activities on consecutive days.
21 Considering added use, but not directly part of the policy, includes consideration of revenue and costs.
22 There is an analysis of the air-tightness of the building.
23 Closed cracks (joints between rotating and fixed parts) of windows and doors.
24 Closed seams (joints between fixed parts) at the junction of window frames on the facades.
25 Closed seams at the connection of the facades and roof to the building walls.
26 Closed seams at the connection of the facades on the ground floor.
27 There are insulated roof ridges.
28 There is insulation of penetrations in the building envelope.
29 The mailbox is either airtight or separated from the building.
30 There is an inventory of energy-using devices in the building.
31 There is a plan to optimize this energy use.
32 Procurement of sustainable energy or on a short-term basis.
33 There is attention to plans and implemented measures.
34 The average insulation status of the building envelope is known.
35 There is a strategy to attain the insulation level set.
36 There is a comparison between alternatives and the existing system.
37 There is a strategy based on this comparison.
38 There is active replacement of non-energy-efficient equipment.
39 There is an exploratory study into the possibility of generating energy on-site.
40 The previously implemented and planned measures are part of the selection of new measures.
41 'Best practice' is part of the process when selecting measures.
42 Sustainability is included in every decision.
43 The combination of measures becomes an 'optimal' system.
44 There is fulfilment of most of the set objectives.
45 There are concrete strategies for the remaining objectives.
46 'Best practice' is part of the building.
47 There is a constant evaluation of measures, combinations and efficiency of the system.

SUBJECT INDEX

AUTHOR INDEX

This index is based on a selection of the references in this book (selection criteria applied: English, recognition of content used, broadly traceable source, closeness to the key subject matter of the book, year of issue). Full listings of authors and titles are given at the end of each chapter.

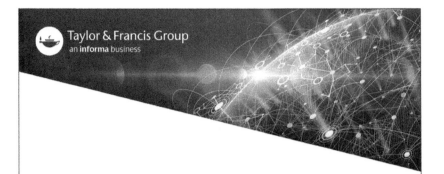